SpringerBriefs in Computer Science

Series Editors

Stan Zdonik
Peng Ning
Shashi Shekhar
Jonathan Katz
Xindong Wu
Lakhmi C. Jain
David Padua
Xuemin Shen
Borko Furht
V. S. Subrahmanian
Martial Hebert
Katsushi Ikeuchi
Bruno Siciliano

For further volumes:
http://www.springer.com/series/10028

Xinxin Liu · Xiaolin Li

Location Privacy Protection
in Mobile Networks

Springer

Xinxin Liu
Xiaolin Li
University of Florida
Gainesville, FL
USA

ISSN 2191-5768   ISSN 2191-5776 (electronic)
ISBN 978-1-4614-9073-9   ISBN 978-1-4614-9074-6 (eBook)
DOI 10.1007/978-1-4614-9074-6
Springer New York Heidelberg Dordrecht London

Library of Congress Control Number: 2013948378

Printed on acid-free paper

Springer is part of Springer Science+Business Media (www.springer.com)

My dearest thanks I dedicate to my father,
for his endless love, constant support,
and the joy he has brought to my life

Xinxin Liu

Preface

During the past a few decades, the world has witnessed significant changes in mobile networking, including both underlying infrastructure as well as communication protocols. With the fast development in miniature hardware and sensing technologies, mobile wireless communication devices are now pervasively used in many human activities. These mobile devices are typically equipped with various types of sensors, which allow them to conveniently collect contextual information of the users and provide personalized service on-the-go. In addition, since mobile devices empower people to access the Internet and stay connected from almost anywhere, information sharing becomes extremely easy. Although these changes greatly benefit people's daily life, they also bring in unprecedented privacy threats, e.g., the potential leakage of the information retrieved or provided by these users, and the potential leakage of the interaction patterns among users and between users and the outer environment.

Privacy preservation policies and technologies have been the topics of heated discussion in the recent years. Among which, location privacy of mobile users is of critical importance, because location information may lead to the exposure of users' personal lives and preferences or even result in physical danger. Research on protecting users location privacy is still at early stage. With the rapid growth of the types of location-based services, more and more privacy issues are waiting to be addressed.

This book is not intended to serve as a comprehensive survey, but is a summary of previous researches the authors have conducted. This brief book can be used by readers who are interested in privacy preserving techniques in mobile networks or by researchers in the field of location privacy as a starting point of their further exploration. We hope this book provides sufficient inspiration for tackling the endless possibilities of privacy threats in mobile networks.

Gainesville, Florida, May 2013 Xinxin Liu
Xiaolin Li

Acknowledgments

This research was supported by the National Science Foundation grants CCF-0953371, CCF-1128805, OCI-0904938, CNS-0709329, and CNS-0916391 at University of Florida and Oklahoma State University.

We would like to express our sincere thanks and appreciation to the people at Springer US, for their generous help throughout the publication preparation process.

Contents

Chapter 1
Introduction

1.1 Mobile Networks: An Overview

The past a few decades have witnessed a surge of growth of mobile wireless communication devices. Starting from 1973 as a marginal feature, mobile communication devices now become a worldwide phenomenon and an indispensable part of people's daily life. Nowadays, over 80 % of the world's population has a mobile phone, and there are more than 5 billion mobile phones worldwide [6]. People use mobile phones to make calls, exchange short messages (SMS), even access the internet. According to a new report by Walker Sands [5], mobile devices now make up about 23.1 % of web traffic in the US alone.

The mobile devices are equipped with radio transmission components, as well as increasingly powerful computing and storage capabilities. Together, they form a wireless communication network called a mobile network. A mobile network can be established through the following ways: (1) relying on a fixed underlying infrastructure; (2) in a self-organizing peer-to-peer fashion; or (3) in a hybrid manner. An example for the infrastructure based mobile network is the cellular network. With the help of cellular infrastructure, mobile phones provide people with the convenience of staying connected with each other from anywhere and at anytime. An example of the self-organized mobile network can be found in military communications. When setting up a fixed infrastructure in enemy territories or inhospitable terrains is impossible, mobile devices carried by soldiers transmit or relay messages whenever two devices are within each other's communication range. The hybrid approach will be taken when a mobile network contains heterogeneous devices with different communication patterns. An example scenario can be a ubiquitous computing environment where both mobile and partially mobile devices are included.

The proliferation of mobile devices and mobile networks not only makes people communicate more easily, but also introduces various new types of information and entertainment services. These services leverage the context information of a user, i.e., the rich information that is collected by the multi-modal sensors equipped on a mobile device, to achieve personalization and improve their service quality. For example,

X. Liu and X. Li, *Location Privacy Protection in Mobile Networks*,
SpringerBriefs in Computer Science, DOI: 10.1007/978-1-4614-9074-6_1,
© The Author(s) 2013

the wide popularity of smartphones empowers people to retrieve information based on their current location or environment. Such context-based services and mobile applications significantly altered people's life styles. Furthermore, other wireless communication devices, such as sensors installed along the roads or buses, and vehicles equipped with GPS and wireless communication capabilities, also provide people with much more accurate and up-to-date information for navigation and traffic prediction. As a result, the ubiquitous mobile devices that are both deeply embedded into the environment as well as tightly coupled with human activities foreshadow the future of pervasive computing.

1.2 Understanding Privacy

Privacy is a concept of global concern [1–3]. In almost every nation, privacy is considered as one of the fundamental rights of a human being [7]. With the fast advances in technologies, privacy is also a concept of increasing importance. Since sharing information becomes extremely easy for users of mobile devices and the Internet, a significant amount of personal information has been put online and shared with friends or even strangers. Such information demonstrates many personal characteristics and preferences of the users, and they are of great interest to advertisers as well as adversarial parties. As a result, more and more privacy breaches are found and reported, and privacy has been the topic of heated discussion for the past a few years.

The term "privacy" is an umbrella term that covers a wide various group of related things. In the broad sense, privacy means the relief from a range of social friction, and it enables people to engage in worthwhile activities in ways that they would otherwise find difficult or impossible [7]. Losing privacy may result in certain things unaccomplished, disruption of reputation, or even physical danger. However, it is well accepted that privacy problem is difficult to articulate precisely and concretely without first giving a specific real world scenario. Therefore, in each chapter of this book, we will spend some time to precisely define the adversary model and privacy protection problem using real world applications.

Generally speaking, privacy vulnerability when using a mobile device comes from three sources: (1) the potential leakage of interactions among users of wireless communication devices, (2) the potential leakage of interaction patterns between a user and the outer environment, (3) the potential leakage of the type of information that these users retrieve or provide. For example, the so journ time of a user at some specific location may indicate the home or work address of that person. Also, information queries about nearby movie theaters that come from a user's mobile communication device may lead to the exposure of the user's current location as well as his/her future activities. Even though the content of the messages provided by or exchanged between mobile devices can be protected through many security mechanisms, the context information, especially the location information of a user can be easily inferred or eaves dropped through the wireless transmission media. Serious consequences may be resulted due to the revelation of a user's context information.

In view of these problems, the topic of this book is the design of privacy preserving techniques in mobile networks to help users obtain the same kinds of services without sacrificing their privacy.

1.3 Scope and Organization of the Book

This book contributes to the investigation of several privacy threats in different mobile network environments, and the design of privacy preservation mechanisms in these mobile networks.

Chapters 2 and 3 study the problem of protecting users' locations and moving trajectories in location based services. Since users who subscribe to location based services need to update their location information periodically using some kind of pseudonym, their information is vulnerable to inferential attack, where an attacker, with the help of collected accidental real identity leakage, unveils the corresponding real identity behind a pseudonym and obtains an extended view of this user's trajectory [4, 8]. To defend against the inferential attack, we propose two approaches. In Chap. 2, we propose a centralized approach using multiple mix zones for generalized privacy protection. We formally model the mix zone location selection problem, and design an efficient algorithm to solve it. In Chap. 3, we design a distributed privacy protection scheme, where each user is able to make decisions based on his/her own privacy requirements.

In Chap. 4, we study the location privacy protection problem for mobile sinks in wireless sensor networks. With the proliferation of mobile devices and the wide deployment of sensor networks, more and more applications suggest using mobile devices as data sinks to collection data from a sensor network. This approach raises the problem of protecting both mobile sinks' and data sources' location information. We propose an energy efficient data collection protocol, named SinkTrail, which preserves location privacy for both data sinks and sensor nodes. The protection effectiveness is analyzed, and the energy efficiency of SinkTrail is validated.

References

1. Amsterdam, A.G.: Perspectives on the fourth amendment. Minn. L. Rev. **58**, 349 (1973)
2. Assembly, U.G.: Universal declaration of human rights. Resolution Adopted by the General Assembly. **10**(12) (1948)
3. Baumer, D.L., Earp, J.B., Poindexter, J.: Internet privacy law: a comparison between the united states and the european union. Comput. Secur. **23**(5), 400–412 (2004)
4. Ma, C.Y., Yau, D.K., Yip, N.K., Rao, N.S.: Privacy vulnerability of published anonymous mobility traces. In: Proceedings of the International Conference on Mobile Computing and Networking (MobiCom) (2010)
5. Quarterly Mobile Traffic Report http://www.walkersands.com/quarterlymobiletraffic (2013)
6. Smartphone users around the world: Statistics and facts. http://www.go-gulf.com/blog/smartphone/ (2011)
7. Solove, D.J.: A taxonomy of privacy. University of Pennsylvania Law Review. **177** 564 (2006)
8. Terrovitis, M., Mamoulis, N.: Privacy preservation in the publication of trajectories. In: Proceedings of the International Conference on Mobile Data Management (MDM), pp. 65–72 (2008)

Chapter 2
Privacy Preservation Using Multiple Mix Zones

In this chapter, we investigate the optimal multiple mix zones placement problem for location privacy preservation. We model the area covered by location-based services as a graph, where all vertices (POIs) are considered as candidates for mix zone deployment. In order to protect mobile users from side information based inferential attacks, we propose to use pairwise vertex association to characterize the linkability of the POIs along a user's trajectory on the map. To achieve maximum privacy preservation, we formulate the optimization problem with the objective of maximizing the overall discontinuity of all possible trajectories on the road network and subject to deployment cost, traffic density, and differentiated privacy priority constraints. For each road segment and intersection, the traffic density effect in terms of entropy is also taken into account. We design three heuristic algorithms corresponding to different traffic scenarios and privacy preservation levels as practical and efficient solutions to the NP-hard optimization problem. Through extensive simulations based on realistic mobile user data traces, we demonstrate that our solution yields satisfactory performance in reducing the success rate of inferential attacks. The mathematical modeling and performance results presented in this chapter offer both theoretical and practical guidance to multiple mix zones placement in mobile networks for protecting users' location privacy.

2.1 Overview

The rapid development of positioning technologies and proliferation of mobile devices have led to the flourish of personalized mobile services based on users' locations, known as Location-Based Services (LBS). Utilizing the underlying network infrastructure, LBS applications are capable of tracking a user's movement and delivering information based on the user's current geographic location. A wide range of mobile LBS applications have been developed to aid people's daily activities, including GPS navigation, social events and friends recommendations, mobile advertising, location-based game [18], etc. According to a recent study conducted by Strategy Analytic, with the increasing consumer demands such as search, maps or navigation, LBS is envisioned to become an over $10-billion-per-year business by year 2016 [21].

Although LBS significantly benefits mobile users, privacy issues arise during the process of collecting, storing, and sharing of users' location information. Users who subscribe to LBS typically have little control over the extent to which their location information is revealed, or with whom the service providers, e.g., smartphone companies and app companies, are sharing this information. Even though users are represented by pseudonyms instead of their real identities in an application, such long-term pseudonym are vulnerable to inferential attacks, where an attacker, with the help of collected accidental real identity leakage, unveils the corresponding real identity behind a pseudonym and obtains an extended view of this user's trajectory [16, 22]. Such real identity leakage, e.g., identity leakage when using credit card at a coffee shop, is known as side information. We use the following example to illustrate the importance of location privacy protection against such attacks. Suppose user Alice, represented by pseudonym Alice', uses LBS at a shopping plaza to query nearby restaurants. She may not mind of others finding out that Alice' corresponds to Alice, and therefore discovering her current location. However, when she later enters a specialized hospital and does not want to share this information with others, revealing the fact that Alice' is Alice becomes a more serious problem (especially when Alice is a well-known public figure). Consequences such as stalking and/or physical crimes may happen due to such revelation of a user's real identity and complete moving trajectory.

Location privacy preservation in mobile environments is challenging for two reasons. First, wireless communications are easy to be intercepted, e.g., an eavesdropper can collect transmitted information of mobile users at certain public place. Besides, since people are publicly observable, context information can easily be obtained from their conversations or behaviors. As a result, partial trajectory information associated with a user's real identity is inevitably exposed to the eavesdropper. Second, the limited resources of mobile devices greatly restrict Privacy-Enhancing Technologies (PET) one could apply and deploy in a wireless network. Consequently, current PET solutions rely on simple schemes to hide the real identity of a mobile user from a passive adversary, rather than complex cryptographic technologies commonly used in wired networks.

To deal with these challenges, a common model for privacy preservation is the mix zone model originally proposed by Beresford and Stajano [3]. A mix zone refers to a service restricted area where mobile users can change their pseudonyms so that the mappings between their old pseudonyms and new pseudonyms are not revealed. A mix zone of k participants therefore becomes a k-anonymization region. For example, Fig. 2.1 shows a mix zone, indicated by the rectangle, deployed at a road intersection. Five users with pseudonyms A-E enter the mix zone from different entrances and exit with a different set of pseudonyms F-J at approximately the same time. The links between old and new pseudonyms are not observable by any outsider. This change effectively "mixes" the identities of all users to achieve privacy

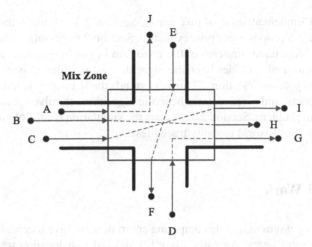

Fig. 2.1 A mix zone example. Rectangular area: a mix zone deployed at a road intersection. Arrows begin or end with dots: perceivable user moving trajectory. Dashed lines: user moving trajectories not perceivable by LBS applications

preservation. To ensure mixing effectiveness, a mix zone typically requires the following conditions to be satisfied:

- At some specific time, there are at least k users inside the mix zone.
- A user enters the mix zone at an entry point, and leaves at an exit point. The probability of transition between any entry point and exit point are equally likely.

Despite the k-anonymity provided by the mix zone concept, deploying a single mix zone in a large area will not provide sufficient protection. In this chapter, we address the problem of optimal multiple mix zones placement to enhance the effectiveness of privacy preservation. Using graph theory, we characterize properties and constraints of the cost-constrained deployment optimization problem, and build a formal mathematical model with the objective of minimizing pairwise information correlation (measured by pairwise node connectivity) over all possible mix zone placement locations. Our contributions in this chapter can be summarized as follows:

- We investigate a new type of attack, i.e., side information based attack, and we propose a new metric to quantify the system's resilience to the side information based attack model [16].
- We present an optimization formulation with realistic deployment constraints to model the multiple mix zones placement problem. Since this formulation is NP-hard, we propose several heuristic algorithms as practical means for finding sub-optimal solutions to the optimization problem.
- We verify the effectiveness of our solution through extensive simulations using real-world mobile user traces.

The rest of this chapter is organized as follows. Section 2.2 summarizes related research in the literature. Section 2.3 presents system model and description of the software level implementation of mix zones. Section 2.4 discusses the adversary model. Section 2.5 proposes our privacy metric. Section 2.6 formulates the mix-zone placement problem under uniform traffic condition in the Integer Linear Programming (ILP) form, and provides heuristic algorithm as an efficient way for finding the sub-optimal solution. Section 2.7 further extends the ILP formulation to incorporate traffic impacts, and discusses two heuristic algorithms to solve the multiple mix zone placement problem. Section 2.8 presents the simulation results of our proposed algorithms using real-world mobility trace files.

2.2 Related Work

Location privacy issues in mobile computing environments have received significant attentions in recent years. An early study [2] showed that location-tracking LBS (locations are tracked by other parties) generates more concerns of privacy leaking than position-aware LBS (device's self-awareness of its current location) for mobile users. Hence, most existing works focus on the location-tracking LBS model and assume the presence of a centralized trusted anonymization server. The most popular technique to achieve the desired level of privacy preservation is to degrade the resolution of location information in a controlled way. This has led to a large number of location perturbation and obfuscation schemes proposed in the last decade. For example, spatial cloaking [10, 11] allows obfuscation of a mobile user's exact location using cloaked spatial areas to meet pre-specified anonymity constraints, such as k-anonymity. However, spatial cloaking may result in a severe degradation of service quality due to the large cloaked area over an extended time period [5]. Therefore, it is in general not suitable to protect privacy in a network-constrained mobile environment such as road networks [23].

An alternative approach for location perturbation and obfuscation is to restrict locating of mobile user position in certain areas, known as the mix zone model [2]. A mix zone often covers a small area, e.g., a road intersection, and allows users to change pseudonyms within the area. Due to its ability to reduce the linkability between identity and trajectory, mix zone deployment over road intersections has gained popularity in vehicular networks. Given the presence of a global passive adversary, Freudiger et al. [8] proposed the CMIX protocol to create cryptographic mix zones at road intersections. Dahl et al. [7] improved the cryptographic approach by fixing the key establishment protocol in CMIX. A more sophisticated protocol, MobiMix [17], improved attack resilience by considering various factors, e.g., traffic density, user movement patterns, etc. All these approaches do not consider the optimal placement of multiple mix zones. Huang et al. [12] proposed to use cascading mix zones. However, their investigation focused on evaluating the QoS implication on real-time applications, rather than protection effectiveness of using multiple mix zones. Shin et al. [19] proposed a request partitioning method to increase the unlinkability of different requests over time. The most related research to our work is presented in [9], where the authors analyzed the optimal placement of multiple mix zones with combinatorial optimization techniques. Our work is distinctive in the following aspects: (1) compared with the flow-based metric used in [9], the accumulated pairwise location associations is more appropriate to capture the global placement effects; (2) based on this metric, our optimal placement strategy is capable of handling a recently emerging side information based attacking model [16] in addition to the simple passive adversary model; and (3) we consider the impact of traffic density at each mix location to enhance the attack resilience.

2.3 System Model

2.3.1 Architecture Overview

The Location-Based Service (LBS) system discussed in this work consists of three major components: Users, Communication/Location Service Infrastructure (CLSI), and third-party Applications Servers, as depicted in Fig. 2.2. Similar system models are commonly found in the literature, e.g., [15, 20]. In such systems, communications happen between users and CLSI, and between CLSI and third-party applications. At any time, the LBS users use their real identities (most likely the identifier associated with their devices) to actively or passively updated their location information to CLSI. CLSI camouflage the users' real identities with pseudonyms, and update the users' location information to third-party applications. Users cannot bypass CLSI and talk directly with third-party applications; otherwise, the applications can trivially obtain the real user identities as well as the user's complete trajectories. A specific application is interested in a set of locations on the map, referred to as Point-Of-Interests (POIs). The physical positions of these POIs may be at road side or road intersections. The application will register these POIs to CLSI. Whenever a user is approaching one of the POIs, the application will be notified by CLSI, and it will send service information through CLSI back to the user. For example, consider a gas-price application designed to provide gas station information tailored to each user's preference around the user's current location in Gainesville, Florida. The application registers all gas stations as POIs at CLSI. Now suppose a subscribed

Fig. 2.2 System model: users, communication/location service infrastructure, and application server

Table 2.1 Notations for privacy preservation using multiple mix zones

Symbol	Definition
$\{\mathbb{P}_i\}$	A set of registered locations within certain range, e.g., POIs in a city where $(i = 1, 2, \ldots, n)$
u_x	User x's pseudonym in the system
v_x	User x's real identity present in the side information
$\mathbb{T}_{u_x}(t_i)$	Per-user time-based function used to describe the location traces collected by an adversary where $(i = 1, 2, \ldots, m; x = 1, 2, \ldots, n)$
t_i	Indicates the time when u_x's location is reported by CLSI
$\mathbb{S}_{v_x}(t_{i'})$	The side information obtained by an adversary where $(i = 1, 2, \ldots, \kappa; x = 1, 2, \ldots, \pi)$

user Alice is driving from home to her work place. The gas-price application will continuously display up-to-date gas prices at neighborhood gas stations along Alice's route. To implement this functionality, the gas-price application will record Alice's preferences and require CLSI to periodically send Alice's location information as callbacks so that the price information at the nearest gas station can be retrieved and reported back to Alice.

The location updates occurred during the service period result in a trajectory file recording a user's footprints. Each entry in the trajectory file is a 3-tuple: ⟨pseudonym, timestamp, location⟩. Based on the trajectory record of Alice, one can estimate the time when Alice arrives at each POI along her complete trajectory. Using a long-term pseudonym is vulnerable to privacy attacks, since one accidental leakage of real identity will result in a user's whole trajectory being compromised. For better privacy preservation, we will employ the mix zone model to break the continuity of a user's trajectory.

The following notations are listed to ease the presentation in the later sections (Table 2.1).

2.3.2 Mix Zone Implementation

Although the theoretical mix zone model discussed in Sect. 2.1 seems to be effective in protecting a user's privacy, these two conditions, i.e., k users exist in a mix zone at some point of time and users have random moving paths, may not be easily satisfied in real world, especially on a road network [17]. A significant amount of research [4, 8, 17] has been devoted to investigating the optimal size and shape of a single mix zone deployment to achieve desired privacy protection. Targeting at vehicular networks, existing single mix zone construction methods are not suitable for the LBS system model, because LBS users can be pedestrians that are not confined to vehicle moving patterns.

Based on the system model presented in Sect. 2.3.1, we propose to establish a mix zone by CLSI at the software level. A mix zone place is selected by CLSI from the set of registered POIs, $\{\mathbb{P}_i\}$, $(i = 1, 2, \ldots, n)$. Once $\mathbb{P}_i$ is chosen as a place for deploying a mix zone, a square shape physical boundary will be set by CLSI. Refer to Fig. 2.1 as an example. The size of the mix zone is determined by CLSI according to the general traffic condition. So that, k users, on average, will present inside the mix zone within a certain period of time. Whenever a user crosses the boundary and enters the mix zone, CLSI will stop all location updates from this user until the user exits from the mix zone. CLSI will give a set of new pseudonyms to the users leaving $\mathbb{P}_i$.

Such a software level mix zone establishment approach has considerable flexibility over physical deployment of mix zones, because the location and the size of the mix zones are not constrained by terrestrial borders and can be easily adjusted. Moreover, the software level mix zone establishment can achieve k-anonymity for general case. Finally, multiple mix zones can be established by CLSI alongside a user's route with much less effort than physical deployment. Consequently, the user's continuous trajectory is broken into a set of discrete segments, where each segment is associated with a unique pseudonym. This causes an adversary to lose the tracking target. Each single mix zone lowers the privacy risk in the user's next trajectory segment.

2.3.3 Mix Zone Effectiveness Measurement

To quantify the protection effectiveness, a commonly used metric for evaluating an adversary's uncertainty in finding out the link between a user's old and new pseudonym in a mix zone model is information entropy given by:

$$H_m = -\sum_u p_u \log p_u, \tag{2.1}$$

where H_m represents the entropy value of a mix zone, and p_u stands for the probability of mapping an old pseudonym to a new pseudonym. According to this metric, we can see that the effectiveness of a mix zone is greatly affected by two factors, the user population and road transit constraint. For example, mix zones deployed at locations with higher traffic density and more outlets have higher entropy than those placed at locations with less or barely no traffic. Therefore, when selecting mix zone locations, traffic density and the number of possible transit paths should be carefully considered.

2.4 Threat Model

In our threat model, we consider CLSI to be trustworthy for two reasons. First, a service provider who operates CLSI generally has no incentive to become adversarial. This is because the service provider who can afford the expensive equipment in CLSI is more likely to be an established major player on the market. The opportunity cost for acting against its customers is too high to afford, e.g., facing expensive law suit and devastating reputation damage. Second, a majority of localization services offered by CLSI rely on message exchange between users and CLSI. In wireless networks, the true identifier of a user's hand-held device is necessary for communication purpose. Therefore, if CLSI is not trustful, we need to consider how to localize a mobile user under the current infrastructure, without exposing any ID information to CLSI. This leads to another set of problems that are out of the scope of this chapter.

The third-party LBS applications are considered not trustworthy. They may directly attack a mobile user's privacy, or secretly sell information to other individuals or organizations. An adversary $\mathbb{A}$ refers to any entity formed by one or more malicious parties (by colluding) who aim at learning the locations associated with mobile users' true identities. The case that $\mathbb{A}$ actively stalks a particular user is considered out of the scope of this chapter. Since an adversary has the complete trajectory profiles camouflaged by pseudonyms, it is often characterized as a global passive eavesdropper, and this type of threat is treated as the major threat in the literature [9].

Besides the trajectory profile, $\mathbb{T}_{u_x}(t_i)$, $(i = 1, 2, \ldots, m; x = 1, 2, \ldots, n)$, a new weapon is brought into sight recently to aid the adversary [16]. Because mobile users are publicly observable, partial trajectory information may be leaked when they travel in public places. For example, information such as "Alice was witnessed to pass by XYZ cafeteria at 3pm", or "Alice used her credit card at XYZ cafeteria at 4pm" becomes valuable auxiliary knowledge to track the mobile target. Such gathered occasional location information forms partial traces of the tracking targets, and becomes side information to $\mathbb{A}$, denoted by $\mathbb{S}_{v_j}(t_{i'})$, $(i = 1, 2, \ldots, \kappa; j = 1, 2, \ldots, \pi)$. Given some side information, the goal of $\mathbb{A}$ is to identify the target mobile user in the trajectory file based on side information matching, and to learn the complete footprints left by the tracking target. For example, in Fig. 2.3, suppose $\mathbb{A}$ obtains user v_j's side information $\mathbb{S}_{v_j}(t_3) = \mathbb{P}_3$, $\mathbb{S}_{v_j}(t_4) = \mathbb{P}_4$, and $\mathbb{S}_{v_j}(t_5) = \mathbb{P}_5$. If $\mathbb{A}$ has already learnt the whole trajectory record from t_1 to t_5 at $\mathbb{P}_1$ through $\mathbb{P}_5$ belonging to some user with pseudonym u_x, by performing side information matching, $\mathbb{A}$ will immediately know that v_j is u_x, and $\mathbb{P}_1$ and $\mathbb{P}_2$ have also been visited by v_j. Therefore, the whole trajectory of v_j is compromised. It must be noted that while the trajectory files contain accurate location records for service purposes, the side information may be noisy or even incorrect. This is because the source of the side information is unreliable, e.g., personal encounter or context inference.

With this established adversary model, we are now able to present our privacy preservation goal as follows: to prevent adversary $\mathbb{A}$ from learning the tracking target's complete trajectory associated with real identity, given partial trajectory may be exposed to $\mathbb{A}$. In the next section, we will present how to quantify this protection goal and build the formal mathematical model to solve the problem.

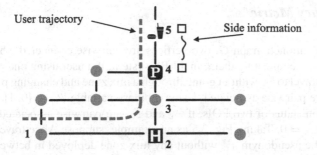

Fig. 2.3 An example of side information and user trajectory in an abstracted POI graph. *Vertices* POIs in a graph. *Edges* road segments connecting POIs. *Dashed line* user trajectory. *Square shapes* connected by *solid curve* side information

2.5 Privacy Preservation Metric

2.5.1 Graph Model

We model the location map with POIs as an undirected graph $G(V, E)$, where V is the set of vertices representing the registered POIs, i.e., $\{\mathbb{P}_i\}$, $i = 1, 2, \ldots, n$, and E is the set of road segments that connect adjacent POIs. In some cases, multiple road segments connecting two POIs may be abstracted as a single edge if there is no POI in-between. An example graph is presented in Fig. 2.4. All vertices in G are considered as potential mix zone deployment locations. A trajectory record belonging to user u_x defines a path consisting of one or a sequence of possibly repeated vertices. Similarly, a piece of side information corresponding to a pseudonym v_x is a portion of some specific trajectory in G. $\mathbb{P}_i$ and index i are used interchangeably to refer to a POI in the following sections.

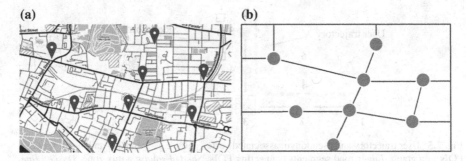

Fig. 2.4 An example of abstracted graph consist of POIs registered by some application servers in an area. The edge between vertices corresponding to the road segments connecting these POIs. **a** An example areamap consists of POIs registered by some application servers, **b** Corresponding graph model consists of POIs represented by vertices

2.5.2 Privacy Metric

In the aforementioned graph G, two vertices are pairwise connected when there is at least one path connecting them. In a LBS system, if a user using one pseudonym from $\mathbb{P}_x$ can travel to $\mathbb{P}_y$ without going through a mix zone and changing pseudonym, $\mathbb{P}_x$ and $\mathbb{P}_y$ are pairwise associated. Using a binary variable $\Psi_{ij} \in \{0, 1\}$ to indicate the association status of two POIs, if $\mathbb{P}_x$ and $\mathbb{P}_y$ are pairwise associated $\Psi_{xy} = 1$; otherwise, $\Psi_{xy} = 0$. Taking Fig. 2.5 as an example, suppose Alice travels from $\mathbb{P}_1$ to $\mathbb{P}_5$ using the pseudonym u_x, without any mix zone deployed in between, we say $\mathbb{P}_1$ and $\mathbb{P}_5$ are pairwise associated. Similarly, $\mathbb{P}_1$ and $\mathbb{P}_4$ are pairwise associated, and $\mathbb{P}_3$ and $\mathbb{P}_4$ are also pairwise associated. An important implication of the pairwise association is that, if u_x appears at $\mathbb{P}_1$, u_x can only appear at locations that are pairwise associated to $\mathbb{P}_1$. Furthermore, once the adversary discovers Alice's pseudonym at $\mathbb{P}_1$, locations that are pairwise associated to $\mathbb{P}_1$ will definitely be compromised if u_x visited them. Given that users change pseudonyms in mix zones, and pseudonyms are unique, placing a mix zone at $\mathbb{P}_3$ will break the pairwise association and protect Alice's future locations, $\mathbb{P}_4$ and $\mathbb{P}_5$, even if her identity is revealed at $\mathbb{P}_1$. We use the total number of pairwise associations in the graph as a privacy metric to quantify the system's privacy preservation level. It is given by:

$$\Phi = \sum_{i,j \in \{\mathbb{P}_i\}} \Psi_{ij}. \qquad (2.2)$$

Our goal in case of the side information based attack is to strategically deploy multiple mix zones so that Φ can be minimized, and the maximum protection level can be achieved for mobile users. Hence, when a user exposes his identity at some point, only limited trajectory can be disclosed by an adversary. Note that there might be multiple paths connecting two vertices in G. The two vertices are dissociated only when all paths in between are blocked by mix zones.

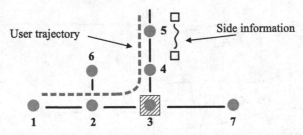

Fig. 2.5 User trajectory and pseudonym associated vertex pairs, e.g., *1* and *2*, *4* and *5*, etc. *Vertices* POIs in a graph. *Edges* road segments connecting POIs. *Shaded square* a mix zone. *Dashed line* user trajectory. *Square shapes* connected by *solid curve* side information

2.6 Uniform Traffic Mix Zone Placement

Given the POI graph model of an area and the pairwise association metric, it is reasonable to argue that the maximum protection level is achieved when mix zones are deployed at all the vertices in G. By doing so, when adversary A discovers Alice's partial trajectory using side information, an immediate pseudonym change can prevent A from learning Alice's future locations. However, deploying mix zones adds certain cost to CLSI, e.g., pseudonym transformation for every user in the mix zone area, saving state information, and informing application servers of newly arrived users. Moreover, mix zones also result in Quality-of-Experience (QoE) degradation perceived by users. When Alice passes by a mix zone area, she might lose services temporarily due to synchronized pseudonym changes. For these reasons, deploying mix zones at all POIs is both expensive and inefficient. We need to strategically plan mix zone placement locations in the system to achieve the maximum location privacy preservation subject to cost and service constraint.

2.6.1 Problem Formulation

In this section, we formulate the multiple mix zone placement problem as an optimization problem, in which the objective function is to minimize the overall number of associated vertex pairs. Since we do not know the probability of the side information exposure at a specific POI, we assume that the side information may include real identity leakages at any POI. Thereafter, our objective function quantifies the global protection effectiveness of deploying multiple mix zones in G.

To fully explore the multiple mix zone placement problem, We first consider mix zone deployment under uniform traffic condition. Although our simplified formulation may not be well-fitted for real world scenario, it is intended to serve as a guideline for the optimal achievable reduction of the total number of pairwise associations. More realistic constraints are considered in later sections.

Cost and service constraints. According to the aforementioned privacy metric, we denote Ψ_{ij} as a binary variable indicating whether there is a path association between vertex i and j in G. Let d_i be another binary variable associated with each vertex i in G. $d_i = 1$ indicates vertex i is selected to be a mix zone; otherwise, $d_i = 0$. Considering cost and service constraints posed on LSI, we limit the number of mix zones to be deployed to be at most K. The constraint is expressed as:

$$\sum_{i \in V} d_i \leq K. \tag{2.3}$$

Graph related constraints. We formulate two graph related constraints to capture the connectivity of the POI graph. The first graph constraint considers two vertices connected by an edge in G. If there is an edge connecting i and j, then there will be a pairwise association between i and j, otherwise, at least one of them should be deployed as a mix zone. That is:

$$\Psi_{ij} + d_i + d_j \geq 1 \quad \forall (i, j) \in E. \tag{2.4}$$

The second graph constraint concerns all vertex triplets. Specifically, the pairwise association is transitive for all vertices in V. If vertex i and j are pairwise associated, and j and k are pairwise associated, then i and j are pairwise associated, meaning there must be some path $i \rightsquigarrow j \rightsquigarrow k$ that a user can travel through without entering into a mix zone. This constraint is described as:

$$\Psi_{ij} + \Psi_{jk} + \Psi_{ki} \neq 2 \quad \forall (i, j, k) \in V. \tag{2.5}$$

In summary, without considering the traffic variety at different roads, we can formulate the optimal mix zone deployment problem as follows:

Minimize $\sum_{i,j \in V} \Psi_{ij}$

Subject to $\Psi_{ij} + d_i + d_j \geq 1 \qquad \forall (i, j) \in E$
$\Psi_{ij} + \Psi_{jv} + \Psi_{vi} \neq 2 \ \forall (i, j, v) \in V$
$\sum_{i \in V} d_i \leq K$
$\Psi_{ij} \in \{0, 1\} \qquad\qquad \forall i, j \in V$
$d_i \in \{0, 1\} \qquad\qquad \forall i \in V$

The ILP formulation of the mix zone placement problem falls into the category of NP-hard problems [6]. A common technique to solve such ILP formulation is to relax the binary constraint $\Psi_{ij}, d_i \in \{0, 1\}$ to a pair of linear constraints $0 \leq \Psi_{ij}, d_i \leq 1$. By doing so, the original NP-hard problem is transformed to a Linear Programming (LP) that is solvable in polynomial time. In general, the optimal solution derived from solving LP does not have all variables either 0 or 1. It cannot be directly used to answer the mix zone placement problem.

2.6.2 Heuristic Algorithm

In this section, we devise a heuristic algorithm as a practical and efficient means to find a suboptimal solution to the mix zone placement. We refer to this heuristic algorithm as Uniform Traffic Mix Zone Placement (UTMP), summarized in Algorithm 1. It provides an estimation of achievable privacy level when no knowledge of traffic patterns is available.

The inputs of UTMP include the abstracted POI graph $G = (V, E)$ and the maximum mix zone number K, which is typically less than the number of vertices in G. The output of UTMP is a set Ω containing selected mix zone locations in G.

In real world, any POI should be reachable from any other POI in the target area. Thus, the area graph is connected without isolated points. Therefore, the total number of possible pairwise connections in such a graph of n vertices is $O(n^2)$. The first step in UTMP is built on the observation that partitioning G into several

---

Algorithm 1: Uniform traffic mix zone placement (UTMP)

---

input : A graph $G = (V, E)$ and K
output: A set Ω of at most K selected mix zone positions
/* --Step #1: Find articulation points-- */
Depth first search for G to find discover time $i.d$ for each vertex;
for *each vertex i in G* **do**
$\quad \lfloor\ i.v \leftarrow \min\{i.d, \min_{backedge\ i \rightarrow w}\{w.d\}\}$;

Initialize articulation points set $\Lambda \leftarrow \varnothing$;
for *each vertex i in G* **do**
$\quad$ **if** $i.v \geq i.d$ **then**
$\quad\quad \lfloor\ \Lambda \leftarrow \Lambda \cup \{i\}$;

$\Omega \leftarrow \Omega \cup \Lambda$;
/* --Step #2: Maximal independent set-- */
Find maximal independent set $\mathbb{I}_{\mathbb{C}_j}$ for each connected component $\mathbb{C}_j$ by iteratively adding
non-adjacent vertices;
$\mathbb{I} \leftarrow \cup \mathbb{I}_{\mathbb{C}_j}$;
$\Omega \leftarrow \Omega \cup \{V \setminus \mathbb{I} \setminus \Lambda\}$;
/* --Step #3: Maintain cost constraint-- */
while $|\Omega| > K$ **do**
$\quad \lfloor$ Find vertex $x \in \Omega$ that contributes the least pairwise associations to $V \setminus \Omega$, and remove
$\quad\quad$ it from Ω;

Return Ω;

Fig. 2.6 An execution snap-shot of our heuristic algorithms. Vertices: POIs in a graph. Edges: road segments connecting POIs. Shaded squares: articulation points. Dash circled vertices: maximal independent set in bottom part of the graph

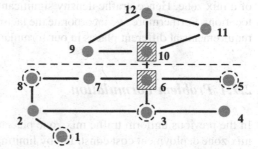

disconnected components is helpful to eliminate the pairwise connections across these components. Hence, we are seeking for vertices whose removal disconnect the graph. Such vertices are typically referred to as articulation points in graph theory, refer to the first step in Algorithm 1. Take the area graph in Fig. 2.6 as an example. Any route from 1 to 9 or from 1 to 12 needs to go through vertices 6 and 10. Therefore, 6 and 10 are articulation points in this graph. If a mix zone is deployed at vertex 6 or 10, a pseudonym appears at any vertex in the bottom part of the graph cannot appear at vertices 9, 12, and 11. Hence, the total number of pairwise associations is reduced.

After G is partitioned into disconnected components, the mix zone deployment in each component is further refined to improve the solution quality. In graph theory, an independent set refers to a set of vertices that are not adjacent to each other. If all vertices that are not in an independent set are selected as mix zones, there will be no pairwise association between the vertices in the independent set, refer to the second step in Algorithm 1. Considering the bottom part of Fig. 2.6 as an example. Circle highlighted vertices, $\{1, 8, 3, 5\}$, form a maximal independent set for the lower part of the graph. If vertices $\{2, 4, 6, 7\}$ are selected as mix zones, a user Alice's pseudonym u_x appears at vertex 1 will not appear at any other vertex in the independent set. As a result, Alice's past and future locations on her trajectory are protected, even though her identity get exposed at vertex 1. Finally, we need to control the number of mix zones to meet the cost and service constraint. At the last step of our algorithm, the maintain cost constraint step in Algorithm 1, we iteratively remove the vertex that introduces the least number of pairwise association increment from the mix zone candidate set selected by previous steps until constraint (2.3) is met. Algorithm 1 summarizes the proposed UTMP algorithm.

2.7 Traffic-Aware Mix Zone Placement

In this section, we extend the uniform traffic multiple mix zone placement formulation to incorporate the impacts of traffic variations at different locations. According to our discussion in Sects. 2.3.2 and 2.3.3, we can see that traffic densities affects both the cost of deploying a mix zone and the privacy preservation effectiveness of a mix zone. Hence, traffic density significantly affects the selection of mix zone locations. Furthermore, we incorporate the factor of different privacy protection level requirements at different places in our formulation.

2.7.1 Problem Formulation

In the previous uniform traffic mix zone placement formulation, we implement the mix zone deployment cost constraint by limiting the maximum total number of mix zones to K. In this section, we employ a more detailed cost model that include both Quality-of-Experience (QoE) degradation constraint of users and mix zone implementation cost constraint of CLSI.

QoE degradation constraints. According to the mix zone implementation described in Sect. 2.3.2, when a user Alice enters a mix zone, she will temporarily lose LBS service due to pseudonym changes. The time period that Alice loses the LBS service is largely related to the mix zone's physical boundary size and its anonymity requirement, e.g. 3-anonymity or 4-anonymity. As a result, different mix zone locations will cause distinct degradation of Quality-of-Experiences (QoE). We denote C_i as the cost of the degradation of QoE at location $\mathbb{P}_\beth$. Once we deploy multiple mix zones in a specific area, the overall QoE degradation can be quantified as $\sum_{i \in V} d_i \times C_i$, which corresponding to the case that a user travels through all the mix zones in the area. To ensure certain level of QoE, we limit the worst case QoE degradation a user can experience as:

$$\sum_{i \in V} d_i \times C_i \leq K_u, \tag{2.6}$$

where K_u is a predefined threshold. This constraint guarantees that if a user travels through all the mix zones in an area, his/her overall QoE degradation will be less than or equal to K_u. The value of C_i can be evaluated according to the size of the mix zone in a particular location. For a specific $\mathbb{P}_\neg$, C_i is a fixed value.

Cost constraints. Since there is both computational and storage cost for CLSI to establish a mix zone, we use $\bar{c}_i$ to represent such cost at $\mathbb{P}_\neg$. The cost of establishing a mix zone at different places will be highly related to the user population at that place, and can be calculated according to the historical traffic density data. We express the cost constraint for establishing mix zones as:

$$\sum_{i \in V} d_i \times \bar{c}_i \leq K_c. \tag{2.7}$$

Traffic related constraints: When traffics are not uniformly distributed around the service coverage area, the difficulty of inferential attack conducted by adversary $\mathbb{A}$ varies significantly. For example, suppose $\mathbb{A}$ observes Alice drives on Main Street at 9:50am, and only one location update belonging to user u_x was recorded in the trajectory profile. Then $\mathbb{A}$ will easily associate u_x with Alice. We use entropy to represent the uncertainty for $\mathbb{A}$ to guess which pseudonym belongs to Alice. It quantifies the inherent attacking resilience for each element in graph G. First, the entropy for a road segment is defined as follows:

$$H_r = -\sum_{u} p_u \log p_u, \tag{2.8}$$

where p_u corresponds to the probability that the identity contained in the side information matches to a particular pseudonym on the road segment.

In addition to road segment entropy, pairwise entropy is useful to describe the uncertainty that an adversary finds out a user has visited both POIs of an associated POI pair. Before defining pairwise entropy, we first clarify the concept of path entropy. A path τ consists of consecutive intermediate vertices between two associated vertices and it has no cycle. The entropy for τ is the expected uncertainty in determining if a user has traveled this path or not. Denote p_{r_i} as the probability that the user's side information is leaked on the ith road segment with road segment entropy H_{r_i}, we have:

$$H_\tau = \sum_{i} H_{r_i} \times p_{r_i}. \tag{2.9}$$

Since there may be multiple paths connecting two vertices, we denote p_{τ_i} as the probability that the user's side information is leaked on the ith path. The pairwise entropy between two vertices is then calculated as:

$$H_p = \sum_i H_{\tau_i} \times p_{\tau_i}. \qquad (2.10)$$

If two vertices have very low pairwise entropy, i.e., they are highly correlated, then we should consider deploying a mix zone to isolate them from other POIs. By doing so, when a user Alice exposes her identity at these two POIs, she can change pseudonym immediately to prevent further location information exposure. A mix zone deployment is considered to be effective only when it satisfies the minimum pairwise entropy requirement. Our proposed model for optimal mix zone placement is traffic-aware because it takes traffic density and entropy into consideration when examining the graph. Specifically, two constraints are defined to ensure the effectiveness of mix zone deployment. First, a mix zone deployed at each vertex on the graph should exceed the predefined entropy threshold ξ_d:

$$(1 - d_i) \times M > \xi_d - e_i \quad \forall i \in V, \qquad (2.11)$$

where M is a very large constant, and e_i is the entropy for location i. In addition to the vertex entropy constraint, we define the following pairwise entropy constraint in our model:

$$(1 - \Psi_{ij}) \times M > \xi_p - \vartheta_{ij} \quad \forall i, j \in V, \qquad (2.12)$$

where ξ_p is a predefined threshold, and ϑ_{ij} is the pairwise entropy for i and j.

Differentiated location priority constraints: While the above formulation incorporates both the cost and traffic related constraints, another realistic requirement need to be addressed. Since different POIs in our proposed graph model corresponds to different physical facilities, e.g., a coffee shop or a hospital, people may pose different privacy requirements at these places. Specifically speaking, people typically demand higher privacy preservation when they are visiting a specialized hospital than when visiting an ordinary coffee shop. Consequently, in order to achieve a better mix zone placement regarding social meanings of the POIs, we employ coarse-grained privacy requirements to differentiate the protection priority for all the candidate areas, which means a place either requires high privacy protection level or requires only ordinary privacy protection level.

A straight forward solution to enhance the privacy protection level of a particular POI is to make it a mix zone. Take a hospital as an example. If user Alice accidentally leaked her real identity before entering a hospital, without changing her pseudonym, an adversary is able to track how long she stayed in the hospital. However, if all user changes pseudonyms at the hospital, it becomes difficult for the adversary to find out whether Alice stayed in the hospital for a certain time period, or she just traveled pass by the hospital.

Denote Γ as the set of POIs that have high privacy requirements, and therefore should be mix zones, we have the constraint

$$d_i \in \{1\} \quad \forall i \in \Gamma. \qquad (2.13)$$

In summary, given the objective function and all constraints, we derive a formal Integer Linear Programming (ILP) formulation for our traffic-aware multiple mix zone placement problem. The complete formulation is described as follows:

$$\text{Minimize } \sum_{i,j \in V} \Psi_{ij}$$

$$
\begin{aligned}
\text{Subject to } \Psi_{ij} + d_i + d_j &\geq 1 & \forall (i,j) &\in E \\
\Psi_{ij} + \Psi_{jv} + \Psi_{vi} &\neq 2 & \forall (i,j,v) &\in V \\
\sum_{i \in V} d_i \times C_i &\leq K_u \\
\sum_{i \in V} d_i \times \bar{c}_i &\leq K_c \\
(1 - d_i) \times M &> \xi_d - e_i & \forall i &\in V \\
(1 - \Psi_{ij}) \times M &> \xi_p - \vartheta_{ij} & \forall i,j &\in V \\
d_i &\in \{1\} & \forall i &\in \Gamma \\
\Psi_{ij} &\in \{0,1\} & \forall i,j &\in V \\
d_i &\in \{0,1\} & \forall i &\in V \setminus \Gamma
\end{aligned}
$$

2.7.2 Heuristic Algorithm

Since the differentiated privacy priority constraint depends highly on the POIs in a specific area, and in some cases, the privacy requirements are generally the same for all the POIs in a particular region, we then propose two heuristic algorithms corresponding to with and without priority constraint, respectively.

2.7.2.1 Non-Uniform Traffic Mix Zone Placement

The first heuristic algorithm aims at solving the multiple mix zone placement problem when CLSI obtains enough historical traffic information over the target area. We name it as **Non-Uniform Traffic Mix Zone Placement** (NUTMP).

As UTMP algorithm, the inputs of NUTMP including the POI graph $G = (V, E)$. NUTMP also requires additional input of entropy and cost information at different POI to take traffic into account, as well as the cost constraints of K_u and K_c. The output of UTMP is a set Ω containing mix zone placement locations in G.

Algorithm 2 summarizes the proposed NUTMP algorithm that further considers the impact of traffic conditions on mix zone deployment effectiveness. Compared with UTMP, NUTMP incorporates two filtering procedures to guarantee the final solution meets the traffic-related constraints (2.11) and (2.12). First, in the articulation point selection step, only those articulation points with entropy values higher than ξ_d are considered as mix zone candidates and put into set Ω. Second, unlike UTMP that selects a maximal independent set as the starting point, in NUTMP, we first choose all vertices that have lower entropy values than ξ_p into a set Ψ so that they cannot be used as mix zones. Then, the vertices that are not articulation points and are not adjacent to any vertex in from Ψ are put into Ψ. The reason for this step is

---

Algorithm 2: Non-uniform traffic mix zone placement (NUTMP)

---

input : A graph $G = (V, E)$, K_u, K_c, mix zone entropies, and entropy matrix for vertex
            pairs
output: A set Ω of selected mix zone positions
/* --Step #1: Find articulation point--   */
Find articulation points set Λ as in Algorithm 1;
Remove the articulation points that have entropy value less than ξ_d from Λ;
/* --Step #2: Non-mix-zone vertices selection--   */
Put all vertices with entropy values less than ξ_d into Ψ;
Select vertices from $V \setminus \Lambda \setminus \Psi$ that are not adjacent to any vertex in Ψ, and put them into Ψ;
$\Omega \leftarrow V \setminus \Psi$;
/* --Step #3: Maintain cost constraint--   */
while *the cost of vertices in Ω exceeds K_u and/or K_c* **do**
    Find vertex $x \in \Omega$ that satisfies the pairwise entropy constraint and incurs the least
    pairwise association increase, and remove it from Ω;
end
Return Ω;

---

similar to the maximal independent set selection in UTMP. By adding non-adjacent vertices to Ψ, no pairwise association is introduced (if all others are mix zones). It is possible that the vertices not qualified to become mix zones are adjacent to each other. If the threshold values are set appropriately, the pairwise entropy constraint should be satisfied in this step. Let Ω become $(V \setminus \Psi)$. By iterating through all mix zone candidates in Ω, we remove those vertices that satisfy the pairwise entropy constraint and incur the least number of pairwise association increment until mix zone cost constraint (2.7) is met.

2.7.2.2 Priority-Aware Non-Uniform Traffic Mix Zone Placement

For a specific area that have POIs with differentiated priority requirements, we devise the Pirority-aware Non-uniform Traffic Mix zone Placement (PNUTMP) algorithm. The inputs of PNUTMP is similar to that of NUTMP wiht an addition of places that must become mix zones. The output of UTMP is a set Ω containing mix zone placement locations in G.

Algorithm 3 provides detailed steps of the proposed PNUTMP algorithm, which enforces the set of critical places identified by CLSI to be mix zones. Specifically, given a set of critical places Γ, e.g., hospitals, which are generally less than K, we apply the first filtering procedure and put them as mix zones. The rest of the steps are similar to the NUTMP algorithm.

Algorithm 3: Priority-aware non-uniform traffic mix zone placement (PNUTMP)

---

input : A graph $G = (V, E)$, K_u, K_c, mix zone entropies, critical place set Γ, and entropy matrix for vertex pairs

output: A set Ω of at most K selected mix zone positions

```
/* --Step #1: Differentiated priority--                          */
```
Select vertices in Γ as mix zones;

if *the cost of vertices in Ω not exceeds K_u and/or K_c* **then**

    Find articulation points set Λ and maximal independent set Ψ as in Algorithm 2;

    $\Omega \leftarrow (V \setminus \Psi) \cup \Gamma$;

```
    /* --Maintain cost constraint--                              */
```
    while *the cost of vertices in Ω exceeds K_u and/or K_c* **do**

        Find vertex $x \in \Omega$ that satisfies the pairwise entropy constraint and incurs the least pairwise association increase, and remove it from Ω;

    end

end

else

    Report no solution can be found;

end

Return Ω;

2.8 Performance Evaluation

2.8.1 Simulation Setup

In this section, we present the simulation results of the proposed UTMP, NUTMP, and PNUTMP algorithms. All algorithms are implemented in C++. Due to the differences in privacy metrics used and problem formulation, it is difficult to conduct direct performance comparison with some existing works, e.g., [9, 12, 19]. To evaluate the solution quality of UTMP, NUTMP, and PNUTMP, we compare the results with the near optimal solution obtained from CPLEX™ [13] using standard solvers, e.g., branch-and-bound algorithm. For trajectory generation, we adopt the real world mobility trace of San Francisco Bay area cabs from CRAWDAD [14]. The partial road map of the same area is abstracted as our input graph. We select 20 POIs from the map covering a diverse location types, e.g., road intersections, hospitals, and bars/coffee shops. For the PNUTMP algorithm, we randomly select 20 % percent of the POIs as the places with high privacy priority.

2.8.2 Mobility Trace Characteristics

The mobility trace [14] of San Francisco Bay Area cabs contains moving trajectories of more than 500 cabs spanning over 20 days time period. In the trace files, each cab is represented by a cab id, and its trajectories are stored in a file named after its

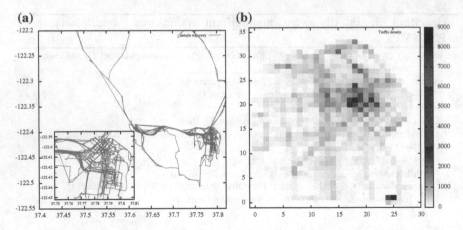

Fig. 2.7 Mobility trace characteristics. **a** Sample trajectory of a user following road network, **b** Spatial histogram of traffic density in the San Francisco Bay area

cab id. The mobility trace of a particular cab includes entries of the form ⟨latitude, longitude, occupancy, time⟩. A sample snapshot of a single cab's moving trajectory over a certain period is shown in Fig. 2.7a. The traffic variations in the Bay Area is also reflected in the cab traces. In Fig. 2.7b, we plotted the spatial histogram showing the cab traffic density per cell within a short period of time.

2.8.3 Protection Effectiveness

First, we compare the solution quality of both UTMP and NUTMP to the solution derived by CPLEX™ (referred to and marked as "near-optimal"). We omit the comparison between PNUTMP and the solution provided by CPLEX for th differentiated priority case, because the places that have high priority are selected as mix zones in both PNUTMP as well as the solution provided by CPLEX, and only the selection of the rest of the vertices may be different. Consequently, it becomes the comparison of NUTMP and near optimal solution obtained from CPLEX. To demonstrate the effectiveness, we also include the simulation results for randomly selected mix zone locations (marked as "random"), and selecting representative mix zones from K evenly partitioned components in G (marked as "even"). The input graph is shown in Fig. 2.9, where all POIs are potential mix zone deployment locations. We evaluate the protection effectiveness for K ranging from 0 to 10. Accordingly, the cost threshold K_u and K_c are calculated as the average cost times K. For the NUTMP algorithm, 20 % of the edges and 10 % of the vertices are randomly selected as low-traffic locations. Their entropy values are drawn from the normal distribution of $\mathbb{N}(1, 0.5)$, and the entropy values for the rest are drawn from the normal distribution of $\mathbb{N}(4, 0.5)$. Fig. 2.8 shows the reduction in total number of pairwise associations when

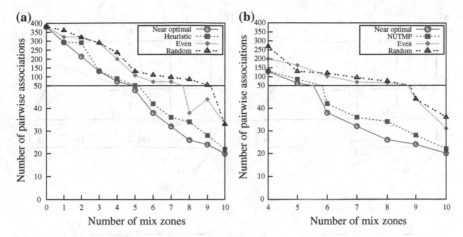

Fig. 2.8 Total number of pairwise associations. **a** Uniform traffic condition, **b** Non-uniform traffic condition

different number of mix zones are deployed in the system. As expected, the number of pairwise associations decreases with the increased number of mix zones in all four methods, under both uniform and non-uniform traffic assumptions. We observe that both UTMP and NUTMP perform very close to the near optimal solution. When the number of the selected mix zones is larger than 4, the average difference of pairwise associations between our heuristic algorithms and the near optimal solutions provided by CPLEX™ is less than 10 %. Because entropy constraints for both vertex and incident edges are taken into account in NUTMP, its outcome is in general different from UTMP. Mostly the value derived from NUTMP is higher than that in UTMP. A possible explanation for this phenomenon is that the ideal locations for minimizing pairwise associations in UTMP may not be qualified in NUTMP because of the traffic-related constraints. Finally, when K becomes larger, the possibility of selection overlapping increases for all methods. Hence, we observe that both "random" and "even" approach performs fairly well when K is large. Figure 2.9 presents an example mix zone selection result to compare the near-optimal solution and our heuristic algorithms. We can see that, the majority of the locations are overlapped. Since the assigned entropy values are low for edges 3 ↔ 5 and 19 ↔ 20, and for vertices 3 and 20, vertex 3 is not selected in Fig. 2.9c, d. Instead, vertex 19 is selected in Fig. 2.9c, d to satisfy the traffic constraint. When the number of mix zones becomes larger, the selected location sets exhibit more overlap. This is the same trend exhibited in Fig. 2.8a, b, where the number of pairwise associations between optimal and heuristic become very close.

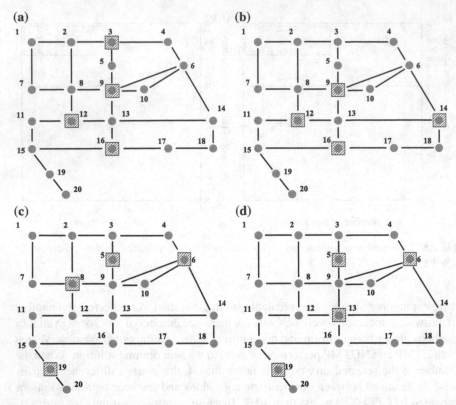

Fig. 2.9 Comparison of mix zone locations between CPLEX's solution and heuristic algorithms. **a** Mix zone deployment by CPLEX under uniform traffic when $k = 4$, **b** Mix zone deployment by UTMP under uniform traffic when $k = 4$, **c** Mix zone deployment by CPLEX under non-uniform traffic when $k = 4$, **d** Mix zone deployment by NUTMP under non-uniform traffic when $k = 4$

2.8.4 Resilience to Inferential Attack

Utilizing the mix zone placement selection results presented in the last section, we conduct another set of simulations to investigate the systems' resilience to side information based attacks. We randomly select 500 partial mobility traces, each has a start and end points, from the San Francisco Bay area cab's mobility traces in CRAW-DAD [14]. Each of them is recorded with a distinct pseudonym. These mobility traces simulate users' trajectories in the input graph. Since the trace file is recorded in ⟨time,coordinates⟩ format, we consider a user stepping onto the corresponding vertex in G, when his trace appears within a certain range of one of the marked POIs. Similarly, the coordinates of a user's trace between two POIs are interpolated and mapped to the closest edge in G. We randomly select some portion of the selected user mobility traces to generate 100 shorter trajectories as side information. Each side information belongs to a particular ID that serves as the real identity of a user. Since real world side information often contains noises [16], we obfuscate the generated side information to better simulate this effect. The maximum likelihood estimation approach for adversary $\mathbb{A}$ is implemented as described in [16] to simulate the side information based inferential attack. An attack is successful if the adversary finds out the corresponding pseudonym used by a user in the side information with high probability. The success rate of an adversary is the ratio of number of successful attacks over total number of attacks. It is worth noting that, due to the uneven distribution of traffic, some of the POIs have are included in more traces than others. As a result, the corresponding entropy values are then calculated. Further more, we intentionally select certain POIs as privacy critical places to test the performance of our PNUTMP algorithm. Figure 2.10 shows the reduction of attack success rate when different number of mix zones are deployed in the target area. According to [16], this type of inferential attack has high success rate when no mix zone is deployed. Using our mix zone deployment algorithms, we observe that the attack success rate can be reduced to over 50 % of original value when 10 mix zones are deployed. Moreover, the difference between our heuristic algorithms and the near optimal solution provided by CPLEX™ is only about 10 % on average. The reason is that, previously, a piece of side information may be able to be matched back to its original mobility trace with high probability. When more mix zones are deployed, this mobility trace may be broken into more pieces of shorter trajectories. It is difficult to find the best match because the side information now faces many possibilities with these broken trajectories under different pseudonyms. In Figure 2.10b, we plotted both the results for both NUTMP and PNUTMP algorithm. From Fig. 2.10 we can see that UTMP, NUTMP, and PNUTMP achieve satisfactory protection effect comparing with the near-optimal solutions, and result in lower attack rate than the other two approaches. Moreover, from Fig. 2.10b we can see that when traffic intensity is considered, better protection effectiveness is achieved. The reason is that when a road segment has high traffic intensity, it is hard to distinguish users on the road with or without the help

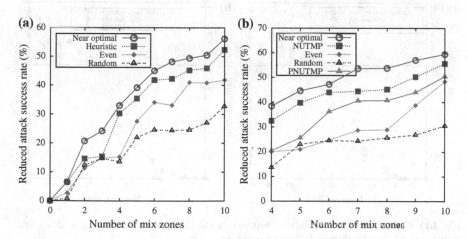

Fig. 2.10 Attack success rate under different traffic and mix zone deployment situations. **a** Uniform traffic condition, **b** Non-uniform traffic condition

of side information. Therefore, the traffic-related constraint provides another level of protection to privacy attack. As for PNUTMP, we fixed 2 POIs as privacy critical places starting from deploying 2 mix zones. Consequently, users passing through the 2 selected POIs will definitely change their pseudonyms, making it more difficult for an adversary to find out who have visited these places. Although the protection effectiveness of PNUTMP may not be as good as NUTMP at first, however, when more numbers of mix zones are deployed, their protecting effectiveness is becoming closer.

2.8.5 Complexity

The complexity of UTMP, NUTMP, and PNUTMP algorithms are contributed by mainly three components. First, the method for finding all articulation points in G is an algorithm suggested by and analyzed in [6]. Its complexity is $O(E)$. Second, finding a maximal independent set by iteratively adding vertices that are not adjacent to current selected vertices requires only linear time in all three heuristic algorithms. The final step in both UTMP, NUTMP, and PNUTMP are similar to the critical node detection algorithm proposed in [1], which has complexity $O(|V|^2|E|)$. As a result, the overall complexity for UTMP and NUTMP are both $O(|V|^2|E|)$.

To validate our complexity analysis of the proposed three heuristic algorithms, we profiled the actual running time of our C++ implementation on various network sizes and edge densities. The experiment environment we used is an Intel Core 2 Duo dual-core processor at 2.66 GHz with 1 GB memory. The results are plotted in Fig. 2.11. As expected from our analysis, the running time of our proposed algorithms is below 15 seconds for a network of size 100. This is much better comparing with

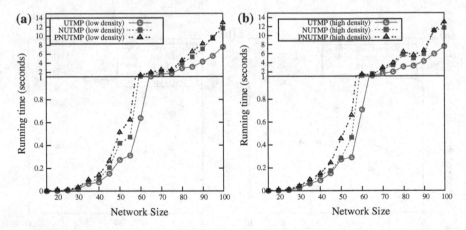

Fig. 2.11 Comparison of execution time between standard ILP solver and the proposed heuristic algorithms. **a** Low edge density, **b** High edge density

over 20 hours' running time for standard ILP solver. As we can see from the figure, increasing either the number of vertices or the number of edges can result in longer running time, which confirms our complexity analysis.

Figure 2.11 provides running time comparison results between the standard ILP solver in CPLEX™ and the proposed heuristic algorithms. The execution environment is the same computer with 3.20 GHz Intel(R) i5 CPU and 4 GB memory. As expected, the heuristic algorithms out-perform the standard ILP solver for the mix zone placement problem.

References

1. Arulselvan, A., Commander, C., Elefteriadou, L., Pardalos, P.: Detecting critical nodes in sparse graphs. Comput. Oper. Res. 36(7), 2193–2200 (2009)
2. Barkhuus, L., Dey, A.: Location-based services for mobile telephony: a study of users' privacy concerns. In: Proceeding of the 9th IFIP TC13 International Conference on Human-Computer Interaction (INTERACT) (2003)
3. Beresford, A., Stajano, F.: Location privacy in pervasive computing. IEEE Pervasive Comput. 2(1), 46–55 (2003)
4. Beresford, A., Stajano, F.: Mix zones: User privacy in location-aware services. In: Proceeding of the 2nd IEEE Annual Conference on Pervasive Computing and Communications Workshops (PERCOMW), pp. 127–131 (2004)
5. Cheng, R., Zhang, Y., Bertino, E., Prabhakar, S.: Preserving user location privacy in mobile data management infrastructures. In: Proceeding of the 6th Workshop on Privacy Enhancing Technologies (PETs), pp. 393–412 (2006)
6. Cormen, T.: Introduction to algorithms. The MIT press, Cambridge (2001)
7. Dahl, M., Delaune, S., Steel, G.: Formal analysis of privacy for vehicular mix-zones. In: Proceeding of the 15th European conference on Research in computer security (ESORICS), pp. 55–70 (2010)
8. Freudiger, J., Raya, M., Félegyházi, M., Papadimitratos, P., Hubaux, J.: Mix-zones for location privacy in vehicular networks. In: Proceeding of the 1st International Workshop on Wireless Networking for Intelligent Transportation Systems (WiN-ITS) (2007)
9. Freudiger, J., Shokri, R., Hubaux, J.P.: On the optimal placement of mix zones. In: Proceeding of the 9th International Symposium on Privacy Enhancing Technologies (PETS), pp. 216–234 (2009)
10. Gedik, B., Liu, L.: Location privacy in mobile systems: A personalized anonymization model. In: Proceeding of the International Conference on Distributed Computing Systems (ICDCS), pp. 620–629 (2005)
11. Gruteser, M., Grunwald, D.: Anonymous usage of location-based services through spatial and temporal cloaking. In: Proceeding of the International Conference on Mobile Systems, Applications and Services (MobiSys), pp. 31–42 (2003)
12. Huang, L., Yamane, H., Matsuura, K., Sezaki, K.: Silent cascade: Enhancing location privacy without communication qos degradation. In: Proceeding of the International Conference on Security in Pervasive Computing (SPC), pp. 165–180 (2006)
13. IBM: IBM ILOG CPLEX optimizer. http://www-01.ibm.com/software/integration/optimization/cplex-optimizer/ (2012)
14. Kotz, D., Henderson, T., Abyzov, I.: CRAWDAD data set dartmouth/campus (v. 2004–12-18). Downloaded from http://www.crawdad.org/dartmouth/campus (2004)
15. Lee, B., Oh, J., Yu, H., Kim, J.: Protecting location privacy using location semantics. In: Proceeding of the 17th ACM International Conference on Knowledge Discovery and Data Mining (SIGKDD), pp. 1289–1297. ACM (2011)

16. Ma, C.Y., Yau, D.K., Yip, N.K., Rao, N.S.: Privacy vulnerability of published anonymous mobility traces. In: Proceeding of the International Conference on Mobile Computing and Networking (MobiCom) (2010)
17. Palanisamy, B., Liu, L.: MobiMix: Protecting location privacy with mix-zones over road networks. In: Proceeding of the International Conference on Data Engineering (ICDE), pp. 494–505 (2011)
18. Schiller, J., Voisard, A.: Location-based services. Morgan Kaufmann, San Francisco (2004)
19. Shin, H., Vaidya, J., Atluri, V., Choi, S.: Ensuring privacy and security for LBS through trajectory partitioning. In: Proceeding of the International Conference on Mobile Data Management (MDM), pp. 224–226. IEEE (2010)
20. Shokri, R., Troncoso, C., Diaz, C., Freudiger, J., Hubaux, J.: Unraveling an old cloak: k-anonymity for location privacy. In: Proceeding of the 9th Annual ACM Workshop on Privacy in the Electronic Society, pp. 115–118. ACM (2010)
21. The 10 Billion Rule: Location, Location, Location. http://www.strategyanalytics.com (2011)
22. Terrovitis, M., Mamoulis, N.: Privacy preservation in the publication of trajectories. In: Proceeding of the International Conference on Mobile Data Management (MDM), pp. 65–72 (2008)
23. Wang, T., Liu, L.: Privacy-aware mobile services over road networks. Proc. of the VLDB Endowment 2(1), 1042–1053 (2009)

Chapter 3
Privacy Preservation Using Game-Theoretic Approach

In this chapter, we propose a distributed approach for LBS privacy protection. In order to protect users from a recently highlighted threat model and achieve k-anonymity, we let distributed mobile LBS users generate dummies according to their own privacy needs when the total number of users in a service area is less than k. From a game theoretic perspective, we identify the strategy space of the autonomous and self-interested users in a typical LBS system, and formulate two Bayesian games for the cases with and without the effect of decision timing. The existence and properties of the Bayesian Nash Equilibria for both models are analyzed. Based on the analysis, we further propose a distributed algorithm to optimize user payoffs. Through simulations using real-world privacy data trace, we justify our theoretical results.

3.1 Overview

The proliferation of Location Based Service (LBS) in mobile hand-held devices has significantly benefited users in many aspects of their daily activities, such as mobile local search, GPS navigation, etc. As LBS evolves, privacy concern becomes more and more important. Protecting location privacy is usually considered as the practice to prevent others from learning a person's past and future locations [3]. To prevent curious eavesdroppers and adversarial third-party applications from learning a user's activities, pseudonyms, instead of real identities, are typically used to camouflage users' location information. Recent works [23, 29] have discovered that even sporadic location information under the protection of pseudonyms is subject to privacy threats. With the aid of side information, an adversary can launch inference attacks to unveil the correlation between the users' real identities and their pseudonyms, and further obtain an extended view of the users' whereabouts to act against their well-being.

The need to protect location privacy is inherent in the LBS system. On the one hand, users need to report sufficiently accurate location information to the LBS server in order to receive high quality services. On the other hand, once the location

X. Liu and X. Li, *Location Privacy Protection in Mobile Networks*,
SpringerBriefs in Computer Science, DOI: 10.1007/978-1-4614-9074-6_3,
© The Author(s) 2013

information is collected, users have no control over how the information will be used by the third-party LBS applications. As a result, the location data is vulnerable to malicious attacks that compromises the privacy of users.

To solve this problem, several location anonymization approaches have been proposed in the literature. Location anonymization refers to the type of approaches that attempt to make a user's location information indistinguishable from a certain number of others. Commonly used techniques include spatial-temporal cloaking or location obfuscation. Because either obscuring or altering a user's geographic position may result in degraded quality of service in LBS, an alternative method is to blend a number of dummy identities with fake location records, known as dummy users or dummies, into normal users' location reports. With the help of the dummies, the users' moving pattern cannot be distinguished from at least $k - 1$ other users, achieving what is known as k-anonymity [28].

The dummy user generation approach is appealing because it effectively achieves k-anonymity without sacrificing LBS quality. However, most existing solutions rely on a trusted Central Authority (CA) [22, 30]. Since it is difficult, and sometimes even impractical to deploy a CA with complete information about all the users in various LBS systems, we tackle the problem of privacy protection in mobile networks from a new angle.

In this chapter, we propose a distributed dummy user generation method to grant users control over their own privacy protections. We let users generate dummies with similar moving patterns as their own according to their diverse privacy needs at different places. Considering the fact that self-interested users may not be sufficiently motivated to generate dummies due to the high cost of dummy generation using mobile devices, we employ game theory to analyze the non-cooperative behavior of LBS users, and identify the equilibrium solutions. To the best of our knowledge, our work is the first to investigate the game-theoretic aspect of distributed dummy generation approach for achieving k-anonymity. We also introduce a novel notion of personalized privacy valuation to differentiate users' diverse privacy needs in time and space. In summary, our contributions are listed as follows:

- To protect against side information aided inference attacks, we propose a distributed approach for achieving k-anonymity by letting users generate dummies according to their privacy needs and valuation at different locations.
- We formally analyze the non-cooperative behaviors of self-interested users from a game theoretic perspective, and formulate two Bayesian game models in both static and timing-aware contexts.
- We analyze the properties of Bayesian Nash Equilibria for both models, and propose a strategy selection algorithm to help users obtain optimized payoffs.
- We conduct simulations based on real-world location privacy data trace, and validate our analytic results.

The rest of the chapter is organized as follows. Section 3.2 summarizes related works in the literature. Sections 3.3 and 3.4 present the system model, the threat model, and define the problem to be addressed in this chapter. Sections 3.5 and 3.6 present the static and the timing-aware game model formulations. Section 3.7

proposes a strategy optimization algorithm. Section 3.8 validates our analytic results through simulations.

3.2 Related Work

Protecting location privacy in LBS has received significant attentions in recent years. To receive LBS, mobile users need to report their location to third-party application servers. Bettini et al. [4] pointed out that a sequence of sporadic location records for a person constitutes a quasi-identifier, i.e., data can be used to identify a person. Several prior works [8, 19, 20] have studied the vulnerability of mobile user privacy to inference attacks. Specifically, inference attack that leverages side information was investigated by Ma et al. [23]. In this chapter, we focus on protecting users' privacy against such side information aided inference attacks.

The concept of k-anonymity was first introduced by Sweeney [28] to measure the effectiveness of location privacy protection. To achieve k-anonymity, most existing works rely on either spatial-temporal cloaking or location obfuscation techniques. Spatial-temporal cloaking [5, 6, 11, 16, 24] refers to the technique that hides an individual's location information by collecting k users' information and send the bounding region as location based query parameters. This approach is generally not suitable for the scenario that users use the same pseudonyms for a certain period of time. On the contrary, location obfuscation alters users' location information, or includes fake or fixed locations rather than the true locations of mobile users as LBS query parameters. Representative approaches such as [2, 7, 18, 22] fall into this category. Both spatial-temporal cloaking and location obfuscation may impair the quality of LBS due to coarse-grained or fake location information. In this chapter, we let users generate dummies with different pseudonyms to achieve location anonymization while preserving LBS quality.

Depending on the system architecture, existing researches can also be classified into centralized or distributed schemes. Centralized schemes, such as [11, 14, 16, 24, 31], rely on a trusted Central Authority (CA) to process queries from users. This type of approach has several drawbacks. First, it is difficult to find or deploy a CA that possesses all user information in various LBS systems. Second, the CA itself may become a bottleneck when the number of LBS queries becomes very large. To overcome these hurdles, we take the distributed approach in this work, and allow autonomous users perform location obfuscation by themselves. The distributed location anonymization approach has also been adopted by [6] and [18]. Kido et al. [18] proposed to let mobile users generate false or dummy locations and send them along with real locations to LBS servers. Targeting at a different threat model, this approach does not require dummy locations to form trajectories similar to real users' trajectories. Chow et al. [6] discussed a peer-to-peer spatial cloaking algorithm, where a mobile user finds a peer group to help achieve k-anonymity when he/she launches queries to LBS provider. This approach works only when there are sufficient number of users present in the targeted area, while in this chapter, we address a different problem that the total number of users is less than k.

Game theory is the study of the interactions among self-interested autonomous individuals. Due to the distributed nature of a mobile environment, game theory is a powerful tool for analyzing privacy protection problems for self-interested mobile users [9, 12, 13, 17, 27]. In this chapter, we investigate a new privacy protection problem that evaluates how k-anonymity can be achieve by letting non-cooperative mobile users generate dummies. Our approach differs from existing solutions in both the adversary model as well as the users' privacy protection strategies.

3.3 Preliminaries

3.3.1 System Model for Location Based Services

A typical LBS system usually consists of the following components: (1) users with mobile devices; (2) communication infrastructure; (3) localization infrastructure; and (4) third-party LBS application servers, as depicted in Fig. 3.1. The mobile devices held by users are capable of utilizing the localization infrastructure, (e.g., GPS system or wireless access points), to pinpoint current geographic positions. In addition, they can establish connections with the communication networks, and report location information to LBS servers. Third-party LBS application servers receive location information from users, search for nearby events in their databases, and then reply the search results back to the users. For the purpose of privacy protection, each user is usually represented by a pseudonym in LBS servers. These pseudonyms are either obtained from an offline central authority or generated in a distributed fashion [10]. The communications occurred during the service period result in a trajectory file recording footprints of users. We define the trajectory function of a user u_i as $\mathbb{L}_{u_i}(t)$. Using long-term pseudonyms are vulnerable to privacy attacks in such applications, because one accidental real identity leakage will result in the whole trajectory, including past and future locations, being compromised (Table 3.1).

The following notations are listed to ease the presentation in the later sections.

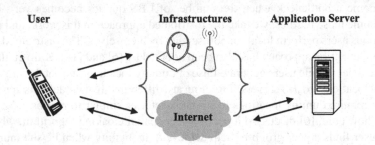

User   **Infrastructures**   **Application Server**

Fig. 3.1 System model: users, localization/communication service infrastructure, and LBS servers

Table 3.1 Notations for privacy preservation using game-theoretic approach

Symbol	Definition
N	Total number of LBS users in current service area
u_i	User i's pseudonym in LBS system
r_i	User i's real identity
$\mathbb{L}_{u_i}(t)$	u_i's trajectory function
$\mathbb{S}_{r_i}(t)$	Side information obtained by an adversary
$\mathscr{P}_i$	Player i
δ_i	Degree of Privacy (DoP)
$\widehat{\delta}$	Maximum DoP for the case of k users
Ψ_i	Valuation of Privacy (VoP) for user u_i
c	Cost of generating $k-1$ dummy users
β_i	Cost-Benefit Ratio (CBR) of u_i
φ	Privacy loss rate
$\Phi(t)$	Privacy loss function
$\tilde{t}$	Earliest dummy user generation time
$\mathscr{F}$	Players' type distribution
f	Probability density function of $\mathscr{F}$
s_i	$\mathscr{P}_i$'s strategy
$s_i(\theta)$	$\mathscr{P}_i$'s strategy as a function of type
$\mathscr{S}$	Strategy profile, i.e., $\mathscr{S} = \{s_1, s_2, \ldots, s_N\}$
$\mathscr{S}_{-i}$	Strategy profile of $\mathscr{P}_i$'s opponents
$\mathscr{U}_i$	$\mathscr{P}_i$'s payoff function

3.3.2 Threat Model

According to the system model, third-party LBS applications are generally considered not trustworthy. After obtaining users' trajectory information, they may directly invade users' privacy, or secretly sell information to other parties. An adversary $\mathbb{A}$ refers to any entity formed by one or more malicious parties, who aim at uncovering the location information associated with mobile users' real identities, and making illegal profit by leveraging this information. Since $\mathbb{A}$ has the complete trajectory profiles camouflaged by pseudonyms, it is often characterized as a global passive eavesdropper. This type of adversary becomes the major targeted threat to deal with in the literature [10].

In addition to the trajectory function obtained by monitoring a user's location reports, $\mathbb{A}$ may also acquire some side information about LBS users. Because mobile users are publicly observable, partial trajectory information may be revealed when they travel in public places, e.g., Alice was witnessed to appear at cafeteria X at 3pm. The side information is represented as $\mathbb{S}_{r_i}(t)$. Although these location disclosures may be sporadic and inaccurate, they are valuable auxiliary information for uncovering users' real identities in LBS systems.

By leveraging the side information, an adversary $\mathbb{A}$ can launch inference attacks. Specifically, $\mathbb{A}$ compares locations and time stamps in the side information and in the

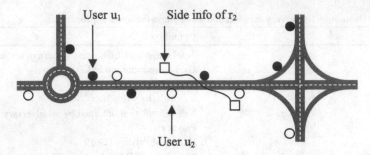

Fig. 3.2 Example of an inference attack based on side information. The trajectories of u_1 and u_2 are obtained from location reports of LBS. Side information associated with u_2's real identity r_2 is obtained by eavesdropping or colluding. Comparing the footprints in these trajectories, an adversary $\mathbb{A}$ is able to reveal the correlation between u_2 and r_2 with high probability. Hence, $\mathbb{A}$ can learn r_2's whole moving trajectory beyond the span of side information

location trajectory file, and identifies the ones that match with the highest probability. An illustrative example is depicted in Fig. 3.2.

The goal of $\mathbb{A}$ is to discover the real identity and pseudonym correlations in the trajectory file based on side information matching, and to uncover the complete footprints associated with users' real identities. The consequence of the inference attacks can be illustrated using the following example. Suppose $\mathbb{A}$ obtains Alice's location reports as follows: $\langle \hat{a}, t_1, \text{``}cafeteria\ X\text{''} \rangle$, $\langle \hat{a}, t_2, \text{``}hospital\ Y\text{''} \rangle$. In addition, $\mathbb{A}$ possesses the side information that *Alice appeared at some place around cafeteria X at time* t_1. Given the situation that only one user presents at the cafeteria X at time t_1, $\mathbb{A}$ may conclude that $\hat{a}$ is the pseudonym of Alice. Further, $\mathbb{A}$ learns the fact that Alice went to hospital Y at time t_2. Note that we do not consider the case that $\mathbb{A}$ actively stalks a particular user in this chapter.

3.3.3 Location Privacy Metric

3.3.3.1 Degree of Privacy (DoP)

Several privacy metrics have been proposed in prior works, among which, the concept of *k-anonymity* is commonly accepted [28]. According to the threat model described in Sect. 3.3.2, *k*-anonymity is defined as the status that r_i is matched with a set of pseudonyms with similar probabilities, where the size of possibly matched pseudonym set is at least k.

In addition to *k*-anonymity, we introduce the concept of Degree of Privacy (DoP) as a quantitative measure for achieved privacy level. The DoP of user r_i at some specific place, denoted as δ_i, is evaluated in terms of an adversary's uncertainty to reveal the correlation of pseudonym to real identity. Leveraging the concept of information entropy, we quantitatively evaluate δ_i as

$$\delta_i = - \sum_{d=1}^{N} p_{(r_i \Leftrightarrow u_d)} \log_2 \left(p_{(r_i \Leftrightarrow u_d)} \right), \tag{3.1}$$

where $p_{(r_i \Leftrightarrow u_d)}$ indicates the probability of matching r_i to u_d. Since we do not make any assumption about the side information, the probability of matching r_i to any pseudonym is equal. Therefore, we consider users in the same area have the same DoP.

3.3.3.2 Valuation of Privacy (VoP)

DoP characterizes the objective measurement of privacy. In order to understand the behaviors of users, we also need to define users' willingness to protect privacy. Depending on a user's context, e.g., current location and time, the willingness to protect privacy may vary significantly. We introduce the concept of Valuation of Privacy (VoP) to quantify the users' subjective perception of privacy. Denote Ψ_i as the VoP of u_i at a specific location. Ψ_i is a function of time t and u_i's trajectory function, i.e., $\Psi_i(\mathbb{L}_{u_i}, t)$. A higher value of Ψ_i indicates a higher privacy appraisement of u_i at a particular place.

3.4 Problem Statement

In this section, we first describe the dummy user generation procedure and clarify the assumptions used in this chapter. Next, we identify the challenges and problems to be addressed in game theoretic analysis.

3.4.1 Dummy User Generation

We consider the scenario when there are less than k users simultaneously present within an LBS service area during a certain period of time. Without employing any privacy protection technique, the achieved DoP for users is less than that of k-anonymity. As a result, according to the threat model, an adversary may have high probability to uncover some of the users' pseudonym to real identity correlations. In order to protect privacy, these users may generate dummy users with random and disposable pseudonyms, and report to the third-party LBS servers. We describe the procedure for dummy user generation as follows. A user first picks a pseudonym by either generating one or selecting one from a pre-loaded pseudonym pool [10, 18]. Next, the user picks some random location within the current area and associates the location with the pseudonym. As long as the user stays within the current service area, these dummy user's locations will keep updated to LBS server

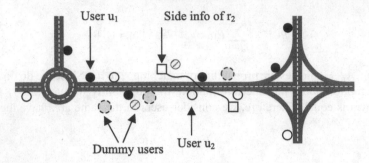

Fig. 3.3 A snapshot showing how dummy users are used to protect against the inference attack. Without dummy user u_2', A may discover the correlation between u_2 and r_2 with high probability. With the newly introduced u_2', the risk of r_2's whole trajectory being revealed is reduced

as if a real user is traveling. This approach effectively enhances privacy for two reasons. First and the foremost, adding dummy users increases the number of possible side information matches, hence effectively decreases matching success possibilities. Second, when an adversary falsely matches a user's current location with some dummy pseudonym, further location information exposure can be avoided due to the disposal of the dummy pseudonym. An illustrative example of how dummy users can help to strengthen the privacy protection effectiveness is illustrated in Fig. 3.3.

3.4.2 Problem Description

Adding dummy users not only enhances privacy protection for the users who actually spend efforts to generate, but also benefits other users within the same area who are passively waiting for others to generate, i.e., free-riders. From the perspective of users, dummy user generation is costly in terms of energy consumption and data communication overhead. Therefore, depending on the perceived VoPs, generating dummy users may not be appealing at all to some of the users, e.g., ordinary people have much lower demands of privacy than celebrities. Given the cost of generating $k - 1$ dummy users represented in c, we define the Cost-Benefit-Ratio (CBR) as:

$$\beta_i = \frac{c}{\Psi_i(\mathbb{L}_{u_i}, t)}. \tag{3.2}$$

Here, we assume β_i falls within the range of $(0, 1)$. When formulating the interactions among self-interested users into a strategy game, the problem boils down to the following. Upon entering into an area of less than k users, whether a user should introduce dummy users to the LBS server to protect his/her privacy, or wait for other users to do so.

We assume that the total number of LBS users in the service area is a common knowledge to all users. There are several ways to accomplish this. For example, in [9, 21], the authors suggested to let users communicate with each other before making any decisions. This method is appropriate for our application scenario because we are targeting at privacy protection against adversarial third-party application servers. For the benefit of their own privacy protection, it is reasonable for the users of the same LBS to communicate (may be using different pseudonyms than the ones used in the application), and obtain the total number of users in current area. Note that in this chapter, we do not consider the case that an adversarial person attacks at this stage.

3.5 Dummy User Generation Game

In this section, we model the scenario that all LBS users within the same service area make dummy user generation decisions simultaneously without being informed of others' choices. We formulate a non-cooperative Bayesian game model to characterize the interactions among self-interested users, and denote this game as the Dummy User Generation (DUG) game. The properties of the Bayesian Nash Equilibria (BNE) with regard to the DUG game are analyzed in details. This game model, albeit simplified, is helpful to gain insights for rational strategy selection in a distributed setting.

3.5.1 Game Model

Player: In the DUG game, the player set $\mathscr{P} = \{\mathscr{P}_i | (i \in \{1, \ldots, N\}\}$ consists of LBS users currently present within the service area. Note that the total number of players is less than k.

Strategy Set: The strategy set of a player refers to all available moves the player is able to take. In the game of DUG, this set includes: (1) *Cooperate (C)*, i.e., to generate $k - 1$ dummy users, or (2) *Defect (D)*, i.e., only report one's own location and wait for others to generate dummy users. Let the strategy for a player $\mathscr{P}_i$ be defined as $s_i \in \{C, D\}$. Regarding the *Cooperate* strategy, a natural question to ask is that how many dummy users should be generated to guarantee k-anonymity. In this work, we enforce the player who chooses the *Cooperate* strategy to generate exactly $k - 1$ dummy users. This is because: (1) generating $k - 1$ users is able to accommodate the worst case when only one user presents in current area; and (2) users are subject to sudden termination of LBS, resulting in a smaller crowd which exposes higher risk of privacy disclosure. Therefore, enforcing *Cooperate* strategy to generate $k - 1$ users will provide guaranteed privacy protection for players. A strategy profile is a collection of the strategies played by all players, i.e. $\mathscr{S} = \{s_1, \ldots, s_n\}$. We use $\mathscr{S}_{-i}$ to indicate the strategy profile of $\mathscr{P}_i$'s opponents.

Payoffs: The payoff of a player $\mathscr{P}_i$ depends on $\mathscr{P}_i$'s own strategy as well as the strategies adopted by $\mathscr{P}_i$'s opponents. Briefly speaking, the payoff of a player equals to the achieved DoP subtracting the dummy user generation cost. In our game model, when $\mathscr{P}_i$ generates $k-1$ dummy users, $\mathscr{P}_i$'s opponents also benefit from this action and accomplish k-anonymity. We use $\widehat{\delta}$ to represent the DoP when k-anonymity is achieved. The value of $\widehat{\delta}$ serves as an upper bound of the achieved privacy level for the k-user case (although users might generate more than k dummies in total, we simply ignore the excessive DoP beyond that of k-anonymity). In this case, a piece of side information can be matched with each of the k pseudonyms with equal possibility, we have

$$\widehat{\delta} = -\sum (1/k) \log_2 (1/k). \tag{3.3}$$

If none of the players chooses to *Cooperate* in the DUG game, the objective of achieving k-anonymity fails. The following payoff function comprehensively covers all cases encountered in the DUG game, where $n_c(\mathscr{S}_{-i})$ denotes the number of cooperating players other than $\mathscr{P}_i$.

$$\mathscr{U}_i(\cdot; \theta_i, s_i, \mathscr{S}_{-i}) = \begin{cases} \widehat{\delta} \times (1 - \beta_i) & s_i = C \\ \widehat{\delta} & s_i = D; n_c(\mathscr{S}_{-i}) \geq 1 \\ 0 & s_i = D; n_c(\mathscr{S}_{-i}) = 0 \end{cases} \tag{3.4}$$

Type: When meeting other players in the area, $\mathscr{P}_i$ only has the information of his/her own VoP. In other words, a player's information about others' payoffs is incomplete. Therefore, the proposed DUG game is a *Bayesian Game*. To deal with the uncertainties inherent in the game, we follow the classical work proposed by Harsanyi [15], where a player named *Nature* is introduced into the game. Each player is assigned a type θ_i sampling independently from some distribution $\mathscr{F}$. The probability density is denoted as f. For type space Θ, we have $\theta \in \Theta$. In Bayesian games, the strategy space, possible types and the probability distribution $\mathscr{F}$ are assumed to be common knowledge. The type of a player $\mathscr{P}_i$'s captures $\mathscr{P}_i$'s private information, i.e., the CBR β_i of $\mathscr{P}_i$ with type θ_i is defined as $\beta_i = \theta_i = c/\Psi_i$.

Since costs are the same for all players in the Bayesian DUG game, the players' strategies are jointly influenced by their VoPs and their beliefs about the VoPs of others. Intuitively, if $\mathscr{P}_i$ believes that others will generate dummy users, *Defect* becomes the natural choice for $\mathscr{P}_i$. We adopt the concept of Best Response to represent the utility maximizing choice of rational players.

Definition 3.1. [Best Response] $\mathscr{P}_i$'s best response $\widehat{s_i}$, given the strategies of other players $\mathscr{S}_{-i}$, is the strategy that maximizes $\mathscr{P}_i$'s payoff. That is

$$\widehat{s_i}(\mathscr{S}_{-i}) = \arg\max_{s_i} \mathscr{U}_i(\theta_i, s_i, \mathscr{S}_{-i}). \tag{3.5}$$

In a complete information game, Nash Equilibrium (NE) captures the steady state of the game, where no player will get better off by unilaterally changing his/her strategy. In the Bayesian DUG game, we are interested in finding an equilibrium state that coheres with classical game theory. Specifically, we have the following definition for the targeted equilibrium state,

Definition 3.2. [Bayesian Nash Equilibrium] A strategy profile $\mathscr{S}^* = \{s_1^*, s_2^*, \ldots, s_n^*\}$ is a Bayesian Nash Equilibrium (BNE) if strategy s_i^* for every player i is the best response that maximizes their *expected payoffs*. That is, given $\mathscr{S}_{-i}$ and players' beliefs about the types of other players θ_{-i}, we have

$$s_i^*(\theta_i) \in \arg\max_{s_i} \sum_{\theta_{-i}} f(\theta_{-i}) \times \mathscr{U}_i(\theta_i, s_i^*, \mathscr{S}_{-i}^*) \; \forall \theta_i \tag{3.6}$$

3.5.2 Bayesian Nash Equilibrium of DUG Game

In order to derive the BNE of the DUG game, we denote $d_i(\theta_i)$ as the probability of player $\mathscr{P}_i$ to choose *Defect* when given type θ_i, and calculate the expected defect probability, η_i, as follows

$$\eta_i = E_{\mathscr{F}}(d_i(\theta_i)) = \int d_i(\theta_i) d\mathscr{F}(\theta_i). \tag{3.7}$$

Given $\mathscr{P}_i$'s type θ_i, $\mathscr{P}_i$'s response to his/her opponents' strategy can lead to one of the following payoffs:

(a) If $\mathscr{P}_i$ chooses *Cooperate*, i.e., $s_i = C$,

$$\begin{aligned} \mathscr{U}_i(C; \theta_i, \mathscr{S}_{-i}, \mathscr{F}) &= \widehat{\delta} \times (1 - \theta_i) \\ &= \widehat{\delta} - \widehat{\delta} \times c/\Psi_i. \end{aligned} \tag{3.8}$$

(b) If $\mathscr{P}_i$ chooses *Defect*,

$$\begin{aligned} \mathscr{U}_i(D; \theta_i, \mathscr{S}_{-i}, \mathscr{F}) &= \widehat{\delta} \times E_{\mathscr{F}}(1 - \prod_{j \neq i} d_j(\theta_j)) \\ &= \widehat{\delta} \times (1 - E_{\mathscr{F}}(\prod_{j \neq i} d_j(\theta_j))) \\ &= \widehat{\delta} - \widehat{\delta} \times \prod_{j \neq i} \eta_j. \end{aligned} \tag{3.9}$$

Combining Eqs. (3.8), (3.9), we summarize the following properties for BNE.

Theorem 3.1. *For a DUG game with N players, each with a type θ_i drawn from some distribution $\mathscr{F}$, where $supp(\mathscr{F}) \subseteq [0, 1]$, the following property holds:*

$$s_i = \begin{cases} C & \text{if } \theta_i < \prod_{j \neq i} \eta_j \\ C/D & \text{if } \theta_i = \prod_{j \neq i} \eta_j \\ D & \text{if } \theta_i > \prod_{j \neq i} \eta_j \end{cases} \tag{3.10}$$

Proof. According to Eqs. (3.8) and (3.9) if $\mathscr{P}_i$ prefers *Cooperate* over other choices, i.e., $\mathscr{U}_i(C; \theta_i, \mathscr{S}_{-i}, \mathscr{F}) > \mathscr{U}_i(D; \theta_i, \mathscr{S}_{-i}, \mathscr{F})$, we have $\theta_i < \prod_{j \neq i} \eta_j$. Other two conditions can be derived in a similar manner. $\square$

Theorem 3.1 indicates that, when the total number of users in the current area increases, it is more likely that a user will choose *Defect*. This conclusion is in accordance with our intuitive judgment. In the extreme case, once there are more than k users in the system, generation of dummy users will become unnecessary.

Theorem 3.2. *For $supp(\mathscr{F}) \subseteq [0, 1]$, there exists N pure-strategy equailibria that exactly one player chooses Cooperate while all other players choose Defect.*

Proof. A pure strategy provides a complete definition of a player's moves. In the context of a DUG game, when $\mathscr{P}_i$'s opponents all choose *Defect*, $\mathscr{P}_i$'s best response is *Cooperate*. Otherwise, $\mathscr{P}_i$ will obtain a payoff of 0 according to Eq. (3.4). As players are symmetric, it does not matter which player is the one that chooses *Cooperate*. Hence, there exists N pure-strategy equilibria. $\square$

Theorem 3.2 proves the existence of N pure-strategy equilibria featuring minimum dummy generation cost when achieving k-anonymity. In fact, these equilibria are also *Pareto efficient*, since no player becomes better off without making others worse off.

Theorem 3.3. *There exists a unique symmetric mixed strategy equilibrium, where the probability of player $\mathscr{P}_i$ chooses Defect is $\eta_i = \widehat{\eta}$, where $i \in \{1, 2, \ldots, N\}$. $\widehat{\eta}$ is evaluated by*

$$\widehat{\eta} = 1 - \mathscr{F}(\widehat{\eta}^{(N-1)}). \tag{3.11}$$

Proof. Symmetric strategy refers to the state that if player $\mathscr{P}_i$ and $\mathscr{P}_j$ have the same type, i.e., $\theta_i = \theta_j$, then $d_i(\theta_i) = d_j(\theta_j)$. Based on Eq. (3.7) and the fixed point theorem [25], we can derive the symmetric case that all players have the same defect probability. Equation (3.11) has a strictly increasing right-hand side and strictly decreasing left-hand side. For $\widehat{\eta} = 0$, the right-hand side is strictly larger than left, and for $\widehat{\eta} = 1$, the right-hand side is strictly smaller than left. When left hand-side equals to right hand-side, only one intersection point exists. Hence, the symmetric mixed strategy is unique. $\square$

A DUG game fails to achieve its privacy protection goal when all the players choose the *Defect* strategy. The conclusion of Theorem 3.3 indicates that when the number of players become large, the possibility of such scenario is very low. However, unlike the pure strategy equilibria discussed in Theorem 3.2, the symmetric mixed strategy equilibrium may result in more than one player to choose the *Cooperate* strategy. In this case, although k-anonymity is achieved, unnecessary cost is also paid.

3.6 Timing-Aware Dummy User Generation Game

In this section, we extend the previously developed DUG game model to incorporate the timing of decisions. Specifically, instead of requiring LBS users that choose *Cooperate* to generate dummy users as soon as they start traveling in the service area, we consider the case where players may intentionally delay their dummy generation and expect others to generate dummy users before they do. Since players make decisions based on their dynamically changing privacy needs, and their beliefs about the dummy user generation time of other players, we call the new game model a Timing-aware DUG game (T-DUG for short). We study the characteristics of the T-DUG game, and analyze the properties of the symmetric BNE for this game.

3.6.1 Extension to DUG Game

The T-DUG game is played whenever several users (less than k) enter into and exit an area at approximately the same time. Entering time and exit time are defined as t_s and t_e, respectively. The value of $t_e - t_s$ corresponds to the time duration within the area. By incorporating the timing factor, we present the extension to the DUG game as follows. Note that this game model is not a repeated DUG game because a player will not choose to generate dummy users twice within a single game.

Strategy set: In the T-DUG game, the strategy of a player refers to the time he/she chooses to generate dummy users. For example, a player $\mathscr{P}_i$ may choose to generate k dummy users at time t_i, $(t_s \leq t_i < t_e)$. This is in accordance with the *Defect* strategy in the DUG game. For the clarity of analysis, we normalize the time duration to the range of $[0, 1]$, where $t_s = 0$ and $t_e = 1$. The value of t_i represents the normalized elapsed time before $\mathscr{P}_i$ generates any dummy user. The strategy profile for the T-DUG game is therefore represented as $\mathscr{S} = \{t_1, t_2, \ldots, t_N\}$. To ease the presentation, we define $\tilde{t} = \min\{t_1, t_2, \ldots, t_N\}$ to be the earliest time of all $t_i \in \mathscr{S}$.

Payoffs: Postponing the generation of dummy users significantly affects the achieved DoP in current area. Imagine that when Alice leaves a crowded shopping mall and enters into a quiet street with very few people around, the chance of Alice's real identity being compromised by some adversary will continuously increase as long as Alice stays on the street. Accordingly, the effective DoP of Alice will continuously decrease after she starts traveling on the street. We introduce the privacy loss rate φ in the current area to capture this effect. We assume that users within the same area share the same privacy loss rate. Before dummy users are generated, the total *privacy loss*, denoted as $\Phi(t)$, can be modeled using a function that is decreasing with time. Due to the page limitation, the discussion of possible modeling functions is omitted. We adopt the following function as an example,

$$\Phi(t) = \exp(-\varphi(t - t_s)) \quad t_s \leq t \leq t_e. \tag{3.12}$$

As a result, the payoff for player $\mathscr{P}_i$ is evaluated as

$$\mathscr{U}_i(t; \theta_i, \mathscr{S}_{-i}, \mathscr{F}) = \begin{cases} \widehat{\delta} \times (\Phi(\widetilde{t}) - \beta_i) & \text{if } t_i = \widetilde{t} \\ \widehat{\delta} \times \Phi(\widetilde{t}) & \text{if } t_i \neq \widetilde{t} \end{cases} \qquad (3.13)$$

Type: The type information for each player is the same as in the DUG game, where type θ_i refers to the hidden information of c/Ψ_i. The same distribution $\mathscr{F}$ with density f is also adopted. In the T-DUG game, we consider that a player's strategy is solely determined by his/her type.

3.6.2 Bayesian Nash Equilibrium of T-DUG Game

We represent $\mathscr{P}_i$'s strategy as $s_i(\theta_i) = t_i$ in T-DUG. Because the strategy space is continuous, we define the expected payoff for player $\mathscr{P}_i$ as:

$$\mathscr{U}_i(t; \theta_i, \mathscr{F}) = E(\widehat{\delta} \times (\Phi(\widetilde{t}) - \theta_i \times I_{[t=\widetilde{t}]})), \qquad (3.14)$$

where I is an indicator function for the condition of $(t = \widetilde{t})$.

According to Eq. (3.14), we have the following result regarding the BNE in T-DUG game.

Theorem 3.4. *The T-DUG game has a unique Bayesian equilibrium, where a player $\mathscr{P}_i$ with type θ_i chooses strategy $s_i(\theta_i)$, defined as*

$$s_i(\theta_i) = \begin{cases} \Phi^{-1}(1 - (N-1)\Omega(\theta_i)) & \theta_i + (N-1)\Omega(\theta_i) < 1, \\ t_e & \text{otherwise} \end{cases} \qquad (3.15)$$

and

$$\Omega(\theta_i) = \int_0^{\theta_i} \frac{x f(x)}{1 - \mathscr{F}(x)} dx. \qquad (3.16)$$

Proof. We employ $G(t)$ to be the distribution of the earliest time $\widetilde{t}$ that any player decides to generate dummy users, and let the density function be $g(t)$. Since a player's strategy is solely determined by his/her type, we have

$$\begin{aligned} G(t) &= Pr(\min_{j \neq i} t_j \leq t) \\ &= 1 - \prod_{j \neq i} Pr(t_j \geq t) \\ &= 1 - \prod_{j \neq i} Pr(s(\theta_j) \geq t) \\ &= 1 - \prod_{j \neq i} Pr(\theta_j \geq s^{-1}(t)) \\ &= 1 - (1 - \mathscr{F}(s^{-1}(t)))^{(N-1)}, \end{aligned} \qquad (3.17)$$

that is

$$G(s(\theta)) = 1 - (1 - \mathscr{F}(\theta))^{(N-1)}. \qquad (3.18)$$

Therefore,

$$g(t) = g(s(\theta)) = (N - 1)(1 - \mathscr{F}(\theta))^{(N-2)} f(\theta)/s'(\theta). \qquad (3.19)$$

Since in a state of equilibrium, players are not incentivized to unilaterally change their strategies, their expected payoffs have reached the maximum value. The expected payoff is thus given by

$$E(\mathscr{U}(t; \theta, G)) = \int_0^t \widehat{\delta} \Phi(x) g(x) dx + \widehat{\delta}(1 - G(t))(\Phi(t) - \theta) \qquad (3.20)$$

By letting

$$\frac{dE(\mathscr{U})}{dt} = \widehat{\delta} \times (\theta g(t) + (1 - G(t))\Phi'(t)) = 0, \qquad (3.21)$$

we have

$$\Phi^{-1}(s(\theta)) = -\theta \times g(s(\theta))/(1 - G(s(\theta))). \qquad (3.22)$$

Substituting $G(t)$ and $g(t)$ with Eqs. (3.17) and (3.19), we have

$$\Phi'(s(\theta)) = -\frac{\theta(N - 1) f(\theta)}{(1 - \mathscr{F}(\theta))s'(\theta)}. \qquad (3.23)$$

By integration over $[0, \theta]$, we get

$$\Phi(s(\theta)) = 1 - (N - 1) \int_0^\theta \frac{x f(x)}{1 - \mathscr{F}(x)} dx. \qquad (3.24)$$

Finally, we have

$$s(\theta) = \Phi^{-1}(1 - (N - 1)\Omega(\theta)). \qquad (3.25)$$

□

3.7 A Distributed Algorithm for Strategy Optimization

In this section, we propose a distributed algorithm for LBS users to choose the best strategy based on local information of VoP and type distribution. As discussed in Sect. 3.4, any user within the current area can serve as a coordinator and initialize the dummy user generation game. The procedure of initialization can be adopted from the Swing protocol proposed in [9, 21]. Similar to the Swing protocol, any user who is in charge of the game coordination (the coordinator) broadcasts an initiation message to other users in proximity. Users who are willing to participate in the game can notify the coordinator by sending back a reply message. After collecting the total number of players, the coordinator broadcasts this information as well as the game type (DUG or T-DUG) to all. The decision of game type can be related to the size of the area. For a small area, DUG game is sufficient. Upon receiving this message from the coordinator, each user is able to proceed to select an optimal strategy based on the theoretical analysis established in Sects. 3.5 and 3.6. For privacy protection, the coordination procedure is conducted without notifying the LBS server. In addition, these messages may be camouflaged using different pseudonyms other than the ones used in the LBS pseudonym pool. The complete description of the distributed algorithm for each participating user is presented in Algorithm 4.

Algorithm 4: LBS User Strategy Selection Algorithm

input : Type distribution $\mathscr{F}$ with density function f
output: Player $\mathscr{P}_i$'s strategy s_i
/* Step #1: Collect game parameters   */
Receive total number of players N and game model: DUG or T-DUG ;
Calculate current player's VoP, Ψ_i ;
Calculate current player's own type, θ_i ;
/* Step #2: Calculate player's strategy   */
if *current game is DUG game* **then**

    $\eta_i \Leftarrow E_{\mathscr{F}}(d_i(\theta_i))$;
    calculate threshold $\prod_{j \neq i} \eta_j$;
    determine the optimal strategy according to Equation (3.10) ;

else

    $\Omega(\theta_i) \Leftarrow \int_0^{(c/\Psi_i)} \frac{xf(x)}{1-\mathscr{F}(x)}dx$;
    if $\theta_i + (N-1)\Omega(\theta_i) < 1$ **then**
      | generate $k-1$ dummy users at time $\Phi^{-1}(1-(N-1)\Omega(\theta_i))$
    else
      | do not generate any dummy user
    end

end

3.8 Performance Evaluation

In this section, we justify our theoretical analysis through simulations leveraging realistic privacy data traces. We use the favorable outcome rate to evaluate the performance of our proposed strategy optimization algorithm for distributed users. Favorable outcome rate is defined as the fraction of the total trials that achieves k-anonymity. In addition, for the T-DUG game, we also investigate the achieved DoP value and the corresponding dummy user generation time for various game size. Our results are evaluated against a random approach, where users randomly decide whether to generate dummy users.

3.8.1 Analysis of Data Trace

Abdesslem et al. organized an experiment in St Andrews and London to collect people's attitudes towards sharing location information with their friends on Facebook. The experiment lasted for about one month, and over 40 voluntary participants were involved. In the experiment, each time when an event took place at a person's smartphone, e.g., checking in at a restaurant or taking a picture, the volunteer was asked whether he/she would like to share this information with his/her Facebook friends. The results are publicly available at [1]. Based on this data trace, we consider the VoPs of mobile users as a function inversely proportional to the fraction of friends with whom the volunteer is willing to share the location information. For example, a user who is willing to share the location information with $\frac{1}{10}$ of his/her Facebook friends has much higher VoP than a user who is willing to share the location information with all friends. Based on this analysis, we plot the empirical distribution of VoP in Fig. 3.4a, and use R [26] to perform distribution fitting. We find out that a Beta distribution with shape parameters $(0.149, 0.109)$ is closest for approximating the empirical distribution of VoPs, which confirms the theoretic estimation about users' privacy attitude distribution in [9].

3.8.2 Results

We conduct simulations for both DUG and T-DUG games. The number of players in a game is increased from 2 to 20 with a step length of 2. The players' type distribution is derived by the standard Jacobian transformation from the distribution of VoP. The values of VoP, Ψ_i, are drawn from $\mathscr{B}(0.149, 0.109)$. For a specific

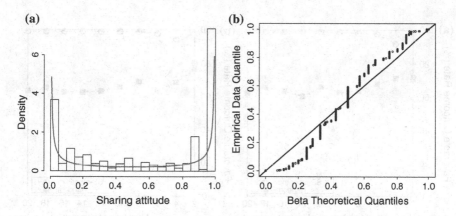

Fig. 3.4 Statistical analysis of the privacy data trace: **a** histogram of the empirical data; **b** O-O plot of the empirical data set and $\mathscr{B}(0.149, 0.109)$

location, e.g., hospital, the type distribution is assumed to be known to all users in a Bayesian game. As described earlier, we compare the simulation results of our algorithm with a random approach based on 1000 simulation runs. In the game of T-DUG, the random approach stands for the process that each player first randomly decides whether to generate dummy users. Once a decision of generation is made, the player will generate the dummy users at a random time between $[t_s, t_e]$.

For each simulation, whenever there is at least one player chooses to generate dummy users, k-anonymity is achieved. The results for the favorable outcome rate of our algorithm and the random approach are depicted in Fig. 3.5. Not surprisingly, for both games, we observe that when there are more than 2 players in the game, our algorithm has a much higher favorable outcome rate than the random approach. This is because the random approach does not take users' privacy preference into account. Conversely, our algorithm maximizes payoffs of users by jointly considering one's attitude towards privacy protection and conjecture about other users' possible strategies. In fact, when the number of players is higher than 4, the favorable outcome rate of our algorithm is almost 100 %. This justifies our theoretical analysis in Theorem 3.3. When the game reaches the symmetric mixed strategy equilibrium, and the number of players becomes large, it becomes almost impossible for all players to choose the *Defect* strategy.

For the T-DUG game, we further investigate the achieved DoPs as well as the earliest time for dummy user generation. The results are plotted in Fig. 3.6. Comparing Fig. 3.6a with Fig. 3.5b, we observe that there is a larger performance gap between our algorithm and the random approach in terms of DoP than that of favorable outcome rate. This is because the counting of favorable outcome rate neglects the effect of dummy user generation time in T-DUG game. In T-DUG game, the sooner dummies are generated, the larger the DoP becomes. This result is further supported by the performance comparison of the earliest dummy generation times in Fig. 3.6b, where the dummy generation time of our algorithm is much earlier than

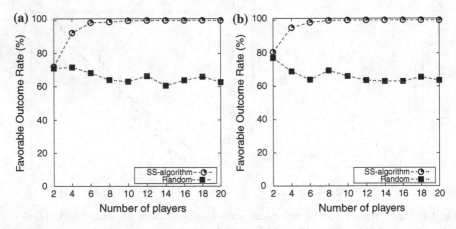

Fig. 3.5 Favorable outcome rate of DUG (**a**) and T-DUG (**b**) game

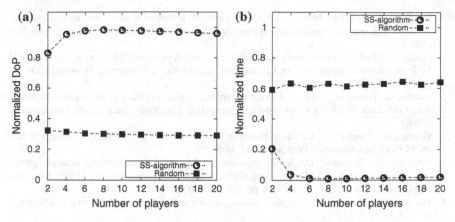

Fig. 3.6 Achieved DoP and timing decisions of players: $\widehat{\delta} = 1$ (normalized) and $\varphi = 2$. **a** Achieved DoP (normalised), **b** Earliest dummy user generation time

the random approach. In addition, in Fig. 3.6b, the earliest dummy generation time of our algorithm approaches zero when the number of players increases. This confirms the theoretical results in Theorem 4.3. An intuitive explanation is that once a player $\mathscr{P}_i$ decides to generate dummies, with the same generation cost, delaying generation incurs higher privacy loss for $\mathscr{P}_i$. Thus, $\mathscr{P}_i$ is more likely to generate dummies as early as possible.

References

1. Abdesslem, F.B., Henderson, T., Parris, I.: CRAWDAD data set st-andrews locshare (v.2011-10-12). http://crawdad.cs.dartmouth.edu/st_andrews/locshare (2011)
2. Ardagna, C., Cremonini, M., Damiani, E., De Capitani di Vimercati, S., Samarati, P.: Location privacy protection through obfuscation-based techniques. Data and Applications Security XXI, pp. 47–60 (2007)
3. Beresford, A., Stajano, F.: Location privacy in pervasive computing. IEEE Pervasive Comput. **2**(1), 46–55 (2003)
4. Bettini, C., Wang, X., Jajodia, S.: Protecting privacy against location-based personal identification. Secure Data Management, pp. 185–199 (2005)
5. Cheng, R., Zhang, Y., Bertino, E., Prabhakar, S.: Preserving user location privacy in mobile data management infrastructures. In: Proceedings of the 6th Workshop on Privacy Enhancing Technologies (PETs), pp. 393–412 (2006)
6. Chow, C.Y., Mokbel, M.F., Liu, X.: A peer-to-peer spatial cloaking algorithm for anonymous location-based service. In: Proceedings of GIS (2006)
7. Duckham, M., Kulik, L.: A formal model of obfuscation and negotiation for location privacy. IEEE Pervasive Comput. (2005)
8. Farkas, C., Jajodia, S.: The inference problem: a survey. ACM SIGKDD Explor. Newsl. **4**(2), 6–11 (2002)
9. Freudiger, J., Manshaei, M., Hubaux, J., Parkes, D.: On non-cooperative location privacy: a game-theoretic analysis. In: Proceedings of the ACM Conference on Computer and Communications Security (CCS), pp. 324–337 (2009)

10. Freudiger, J., Shokri, R., Hubaux, J.P.: On the optimal placement of mix zones. In: Proceedings of the 9th International Symposium on Privacy Enhancing Technologies (PETS), pp. 216–234 (2009)
11. Gedik, B., Liu, L.: Location privacy in mobile systems: A personalized anonymization model. In: Proceedings of the International Conference on Distributed Computing Systems (ICDCS), pp. 620–629 (2005)
12. Gianini, G., Damiani, E.: A game-theoretical approach to data-privacy protection from context-based inference attacks: a location-privacy protection case study. Secure Data Management (2008)
13. Gianini, G., Damiani, E.: Cloaking games in location based services. In: Proceedings of the ACM Workshop on Secure Web Services (2008)
14. Gruteser, M., Grunwald, D.: Anonymous usage of location-based services through spatial and temporal cloaking. In: Proceedings of the International Conference on Mobile Systems, Applications and Services (MobiSys), pp. 31–42 (2003)
15. Harsanyi, J.: Games with incomplete information played by "bayesian" players, i-iii. Manag. Sci. **14**, 159–182 (1967)
16. Hoh, B., Gruteser, M.: Protecting location privacy through path confusion. In: Proceedings of SecureComm (2005)
17. Karimi Adl, R., Askari, M., Barker, K., Safavi-Naini, R.: Privacy consensus in anonymization systems via game theory. Data and Applications Security and Privacy XXVI. **7371**, 74–89 (2012)
18. Kido, H., Yanagisawa, Y., Satoh, T.: An anonymous communication technique using dummies for location-based services. In: Proceedings of the International Conference on Pervasive Services (ICPS) (2005)
19. Krumm, J.: A survey of computational location privacy. Pers. Ubiquitous Comput. **13**, 391–399 (2009)
20. Krumm, J.: Inference attacks on location tracks. IEEE Pervasive Comput. 127–143 (2007)
21. Li, M., Sampigethaya, K., Huang, L., Poovendran, R.: Swing & swap: user-centric approaches towards maximizing location privacy. In: Proceedings of the ACM Workshop on Privacy in Electronic Society (2006)
22. Lu, H., Jensen, C.S., Yiu, M.L.: Pad: privacy-area aware, dummy-based location privacy in mobile services. In: Proceedings of MobiDE (2008)
23. Ma, C.Y., Yau, D.K., Yip, N.K., Rao, N.S.: Privacy vulnerability of published anonymous mobility traces. In: Proceedings of the International Conference on Mobile Computing and Networking (MobiCom) (2010)
24. Mokbel, M.F., Chow, C.Y., Aref, W.G.: The new casper: query processing for location services without compromising privacy. In: Proceedings of the 32nd International Conference on Very Large Data Bases (VLDB), pp. 763–774. VLDB Endowment (2006)
25. Osborne, M.: An introduction to game theory. Oxford University Press, New York (2004)
26. Project, T.R.: R software. http://www.r-project.org/ (2012)
27. Shokri, R., Theodorakopoulos, G., Troncoso, C., Hubaux, J., Le Boudec, J.: Protecting location privacy: Optimal strategy against localization attacks. In: Proceedings of the ACM Conference on Computer and Communications Security (CCS) (2012)
28. Sweeney, L.: k-anonymity: a model for protecting privacy. Int. J. Uncertainty, Fuzziness Knowl. based Syst. **10**, 557–570 (2002)
29. Terrovitis, M., Mamoulis, N.: Privacy preservation in the publication of trajectories. In: Proceedings of the International Conference on Mobile Data Management (MDM), pp. 65–72 (2008)
30. Tran, M., Echizen, I., Duong, A.: Binomial-mix-based location anonymizer system with global dummy generation to preserve user location privacy in location-based services. In: Proceedings of ARES (2010)
31. Wang, T., Liu, L.: Privacy-aware mobile services over road networks. In: Proceedings of the VLDB Endowment **2**(1), 1042–1053 (2009)

Chapter 4
Privacy Preservation Using Logical Coordinates

When mobile devices are used as mobile sinks for collecting information from a deployed sensor network, the location privacy of both mobile sinks and data sources becomes very important for some applications. We present the SinkTrail and its improved version, SinkTrail-S protocol, two low-complexity, proactive data reporting protocols for privacy preserving and energy-efficient data gathering. SinkTrail uses logical coordinates for location privacy protection and to establish data reporting routes. In addition, SinkTrail is capable of accommodating multiple mobile sinks simultaneously through multiple logical coordinate spaces. It possesses desired features of geographical routing without requiring GPS devices or extra landmarks installed. SinkTrail is capable of adapting to various sensor field shapes and different moving patterns of mobile sinks. We systematically analyze energy consumptions of SinkTrail and other representative approaches and validate our analysis through extensive simulations. The results demonstrate that SinkTrail preserves location privacy as well as effectively reduces overall energy consumption. The impact of various design parameters used in SinkTrail and SinkTrail-S are investigated to provide guidance for implementation.

4.1 Overview

People are now living in a increasingly digitized world. The advances in miniature hardware and communication technologies have made the ubiquitous availability of computing/communication infrastructure a reality. Under the paradigm of ubiquitous computing, Wireless Sensor Networks (WSNs) have enabled a wide spectrum of applications through networked low-cost low-power sensor nodes, e.g., habitat monitoring [22], precision agriculture [18], and forest fire detection [38]. In these applications, the sensor nodes constantly collect data from their immediate vicinity, and report back their readings to a data sink either actively or passively. These sensor networks are deeply embedded into the environment and operate under few human interventions.

X. Liu and X. Li, *Location Privacy Protection in Mobile Networks*,
SpringerBriefs in Computer Science, DOI: 10.1007/978-1-4614-9074-6_4,
© The Author(s) 2013

Although a significant amount of research has been devoted to improve the energy efficiency of WSNs and prolong the network lifetime, privacy issues in WSNs have not been thoroughly addressed. The problem of privacy protection in WSNs has many aspects. First, as sensor nodes are deployed to extract information from their proximity, the data collected are highly correlated with the specific region. Therefore, for sensitive data, such as the discovery of some scarce animals, the source of the data should be protected against the adversarial hunters. Second, the data sink may become the target of an adversary. As data aggregation and analysis are performed at data sinks, once the location of a data sink is compromised, serious consequences, e.g., physical damage, may be resulted. Moreover, many sensor network applications allow people carrying their mobile devices to walk around and collect sensor readings from the surrounding environment in real time. In such a scenario, privacy becomes of great importance when people's locations and activities are involved. Finally, because sensor nodes typically have very limited battery life, energy saving is of paramount importance in the design of sensor network protocols.

To address these problems, we propose an energy efficient, mobile sink based data collection protocol that hides the location information of both sensor nodes and the mobile sinks, (e.g., people carrying mobile devices). Using our protocol, mobile sinks move continuously in an area with relatively low speed, and gather data on the fly. Control messages are broadcast at certain points in much lower frequency than ordinarily required in existing data gathering protocols. These sojourn positions are viewed as "footprints" of a mobile sink. Considering each footprint as a virtual landmark, a sensor node can conveniently identify its hop count distances to these landmarks. These hop count distances combined represent the sensor node's coordinate in the logical coordinate space constructed by the mobile sink. Each sensor node greedily selects next hop with the shortest logical distance to the mobile sink. In this way, no actual geographic location information is leaked during the data collection process, and location privacy of both sensor nodes and mobile sinks are protected.

Our contributions in this chapter are manifold. (1) We solve the privacy protection problem for mobile sinks based data collection scenario by utilizing a unique logical coordinate system. As a result, the location information of the data source, i.e., sensor nodes, and the mobile sinks are effectively protected. (2) We design a novel low-complexity geographic routing protocol based on logical coordinates. Our routing protocol significantly reduces average route length and cuts down total energy consumption. (3) We conduct extensive comparison studies and simulations with popular existing solutions to demonstrate the effectiveness of our approach.

The rest of this chapter is organized as follows. Section 4.2 presents related work. Section 4.3 discusses network and adversary models. Section 4.4 introduces detailed protocol design. Section 4.5 presents analytic and simulation results, and demonstrates the effectiveness of our approach in terms of privacy protection and energy efficiency.

4.2 Related Work

Protecting location privacy in sensor networks has been a heated topic recently [1, 9, 30, 34]. Many approaches have been proposed in this area to preserve either sensor nodes' location privacy or data sinks' location privacy. Even though the classic security approaches, e.g., encryption of data packets, is able to achieve confidentiality and integrity, location information is vulnerable to traffic analysis. To make things worse, with the extremely limited on-board energy resources of a sensor node, energy-efficiency is also an essential requirement for all types of protocol designs in sensor networks.

4.2.1 Data Source Privacy

For many sensor network applications that aim at monitoring or tracing a specific target, the main purpose of protecting data source's location privacy is to avoid the leakage of the target location to eavesdroppers. According to the packet generation pattern, they can be generally classified into the following categories. (1) The flooding technique, e.g., [27], suggests each data source send packets through multiple paths to avoid identification of target location through packet trace back. (2) The random walk technique, e.g., [13], obfuscates the forwarding path to avoid source identification. (3) The cyclic entrapment method, e.g., [26], makes packets travel in a cyclic pattern. Besides, there are some research work proposes to protect source location privacy by generating fake packets from time to time to confuse the adversarial eavesdroppers [13, 35]. Although all these techniques can work for some applications, they all require extra efforts from sensor nodes in generating or forwarding unnecessary packets. As a result, they are not very energy efficient. Our proposed approach is specifically designed to achieve source protection in an energy efficient manner.

4.2.2 Data Sink Privacy

Similar to the data source privacy protection approaches, methods designed to protect data sinks' location also focus on altering the traffic patterns. Multi-path routing, multiple-parent routing, and a controlled random walk scheme are introduced in [5] and [6]. While in [12], redundant hops and fake packets are added to increase the difficulty of identifying sink location. In [23] and [24], the sink location protection against a global eavesdropper is studied, and sink location simulation and backbone flooding approaches are proposed. All these existing approaches achieve protection by sacrificing energy efficiency of routing in a sensor network. In addition, when the data sink is a mobile node, techniques proposed for static network structure will no longer work. Unlike the existing approaches, we study the protection of a mobile sink's location privacy in this chapter.

4.2.3 Energy Efficient Routing

For mobile sink based data collection, broadcasting a mobile sink's current location to the whole network is the most natural solution to track a moving mobile sink. This type of approach is sink-oriented and some early research efforts, e.g., [3, 11, 37], have demonstrated its effectiveness in collecting a small amount of data from the network. Several mechanisms have been suggested to reduce control overheads. The TTDD protocol, proposed in [36], constructed a two-tier data dissemination structure in advance to enable fast data forwarding. In [10] and [28], a spatial-temporal multicast protocol is proposed to establish a delivery zone ahead of mobile sink's arrival. Fodor et al. [7] lowered communication overheads by proposing a restricted flooding method. Luo et al. [20] proposed that a mobile sink should move following a circle trail in deployed sensor field to maximize data gathering efficiency. One big problem of the multicasting methods lies in its flooding nature. Moreover, these papers either assume that mobile sinks move at a fixed velocity and fixed direction, or follow a fixed moving pattern, which largely confines their application. The Sink-Trail protocol with message suppression minimizes the flooding effect of control messages without confining a mobile sink's movement, thus is more attractive in real-world deployment. Another solution utilizes opportunistic data reporting. For instance, in [31] the authors studied data collection performance when a mobile sink presents at random places in the network. The method relies heavily on network topology and density, and suffers scalability issues when all data packets need to be forwarded in the network.

Another category of methods, called mobile element scheduling (MES) algorithms [4, 21, 32, 33, 39–41], considered controlled mobile sink mobility and advanced planning of mobile sink's moving path. Ma et al. [21] focused on minimizing the length of each data gathering tour by intentionally controlling the mobile sink's movement to query every sensor node in the network. When data sampling rates in the network are heterogeneous, scheduling mobile sinks to visit hot-spots of the sensor network becomes helpful. Example algorithms can be found in [4, 32, 33]. Although the MES methods effectively reduce data transmission costs, they require a mobile sink to cover every node in the sensor field, which makes it hard to accommodate to large-scale and introduces high latency in data gathering. Even worse, finding an optimal data gathering tour in general is itself an NP-hard problem [19, 21], and constrained access areas or obstacles in the deployed field pose more complexity. Unlike MES algorithms, SinkTrail, with almost no constraint on the moving trajectory of mobile sinks, achieves much more flexibility to adapt to dynamically changing field situations while still maintains low communication overheads.

SinkTrail uses sink location prediction and selects data reporting routes in a greedy manner. The authors in [16] used sequential Monte Carlo theory to predict sink locations to enhance data reporting. SinkTrail employs a different prediction technique that has much lower complexity. Moreover, SinkTrail does not rely on the assumption of location-aware sensor nodes, which could be impractical and lead to location information leakage. The routing protocol of SinkTrail is inspired by recent research

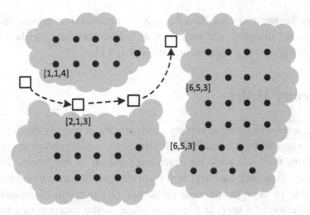

Fig. 4.1 Data gathering with one mobile sink: *square shapes* indicate the mobile sink's trail points, and sensor nodes maintain trail references as logical coordinates. *Shaded areas* stand for different parts of the sensor networks

on logical coordinate routing [2, 8, 25, 29]. SinkTrail adopts a vector representation and uses past locations of the mobile sink as virtual landmarks. To the best of our knowledge, we are the first to associate a mobile sink's "footprints" left at moving path with routing algorithm construction. The vector form coordinates, called trail references, are used to guide data reporting without knowledge of the physical locations and velocity of the mobile sink.

4.3 Network and Adversary Model

4.3.1 Network Model

We consider a large scale, uniformly distributed sensor network $\mathbb{N}$ deployed in an outdoor area. For analysis purpose, we assume all sensor nodes are homogeneous, although this assumption can be relaxed with small modification of the underlying communication protocol. Figure 4.1 shows an example deployment. Nodes in the network communicate with each other via radio links. We assume the whole sensor network is connected, which is achieved by deploying sensors densely. We also assume sensor nodes are awake when a mobile sink wants to start the data gathering process (achievable by synchronized schedule or a short "wake up" message). Mobile sinks, e.g., tourists of an area or rangers of a park, travel in the area and collect data from deployed sensor nodes. These mobile sinks have radios and processors to communicate with sensor nodes and process sensed data. A data gathering process starts from the time mobile sinks issue a start message and terminates when: either (1) enough data are collected (measured by a user defined threshold); or (2) there are no more data reports in a certain period.

Since many security mechanisms have been proposed to ensure data contents' confidentiality and integrity, we assume that the sensed data can be encrypted by sensor nodes using existing approaches, e.g., [14, 15].

4.3.2 Adversary Model

In this chapter, we focus on an adversary with only local view of the deployed sensor network. For example, by randomly deploying some snooping device in an area, the adversary obtains the local view via eavesdropping on its neighbors. The adversary can utilize the routing path and next hop information that are carried in the packets, but it is not able to decrypt message contents. The purpose of an adversary is to discover the location of a specific events as while as the location of a mobile sink.

4.3.3 Privacy Protection Goal

Since our targeted network utilizes mobile sinks for data collection and query, sink nodes may appear at anywhere inside the deployed area. Without a global view of the traffic patterns, an adversary can only rely on routing information carried in each packet to find out sink and source locations. The goal of privacy protection against such a local eavesdropper is to hide the location information in exchanged packets. By establishing a logical coordinate system, the goal of protecting location information and energy efficient routing are jointly achieved.

4.4 SinkTrail Protocol Design

The SinkTrail protocol is proposed for sensor nodes to proactively report their data back to one of the mobile sinks in an energy efficient and privacy preserving manner. To illustrate the data gathering procedure clearly, we first consider the scenario where there is only one mobile sink in $\mathbb{N}$. The multiple mobile sinks scenario is discussed in Sect. 4.4.2

4.4.1 SinkTrail Protocol with One Mobile Sink

During the data gathering process, the mobile sink moves around in $\mathbb{N}$ with relatively low speed, and keeps listening for data report packets. It stops at some places for a very short time, broadcasts a message to the whole network, and moves on to another place. We call these places *"Trail Points"*, and these messages *"Trail Messages"*. Example trail points are shown in Fig. 4.1. Let $\bar{\tau}$ be the average transmission range. Apparently two adjacent trail points should be separated by a distance longer than $\bar{\tau}$, otherwise, the hop count information won't be significantly different. To facilitate the tracking of a mobile sink, we assume that the distances between any two consecutive

---

Algorithm 5: Mobile Sink's Strategy

```
/* ------Initialization------                                    */
msg.seqN ← 0 ;
msg.hopC ← 0 ;
Announces step size parameter K ;
/* -----Moving strategies------                                  */
while Not get enough data or Not timeout do
    Move to next trail point ;
    msg.seqN ← msg.seqN +1 ;
    Stop for a very short time to broadcast trail message ;
    Concurrently listen for data report packets ;
end
End data gathering process and exit ;
```

Table 4.1 Notations for privacy preservation using using logical coordinates

Symbol	Definition
n_i	Sensor node i
N	Total number of sensor nodes in $\mathbb{N}$
S	Mobile sink
M	Number of mobile sinks
v_i	Trail reference of node i
$e_i{}^j$	the jth element in v_i
d_v	Trail reference size, $d_v = \|v\|$
b	Average number of neighbors of each node
λ	The most recent message sequence number
π_i	The ith trail point of S
Π	The collection of trail points
D_π	Total number of trail points
K	Step size parameter for one move (A step of K hop counts is $K\bar{\tau}$)
T_i	Timer duration of node i

trail points are same (or similar), denoted as $K\bar{\tau}, K \geq 1$. However, distribution of these trail points doesn't necessarily follow any pattern. A trail message from a mobile sink contains a sequence number ($msg.seqN$) and a hop count ($msg.hopC$) to the sink. The time interval between a mobile sink stops at one trail point and arrives at the next trail point is called one *"move"*. There are multiple moves during a data gathering round. The tasks of a mobile sink is summarized in Algorithm 5.

In the SinkTrail algorithm, we use vectors called *"Trail References"* to represent logical coordinates in a network. The trail reference maintained by each node is used as a location indicator for packet forwarding. All trail references are of the same size. Notations used throughout the protocol description are listed in Table 4.1.

The data reporting procedure consists mainly two phases. The first phase is called logical coordinate space construction. During this phase, sensor nodes update their trail references corresponding to the mobile sink's trail messages. After d_v hop counts

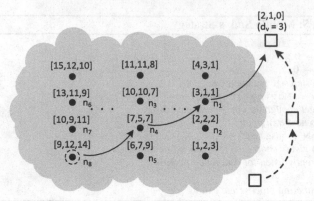

Fig. 4.2 Example execution snapshot of SinkTrail: *square shapes* indicate trail points and the mobile sink's moving path

have been collected, a sensor node enters the greedy forwarding phase, where it decide how to report data packets to the mobile sink.

4.4.1.1 Logical Coordinate Space Construction

At beginning, all sensor nodes' trail references are initialized to $[-1, -1, \ldots, -1]$ of size d_v. A special variable λ that is used to track the latest message sequence number is also set to -1. After the mobile sink S enters the field, it randomly select a place as its first trail point π_1, and broadcasts a trail message to all the sensor nodes in $\mathbb{N}$. The trail message, $<msg.seqN,msg.hopC>$, is set to $< 1, 0 >$, indicating that this is the first trail message from trail point one, and the hop count to S is zero.

The nodes nearest to S will be the first ones to hear this message. By comparing with λ, if this is a new message, then λ will be updated by the new sequence number. And node n_i's trail reference v_i is updated as follows. First, every element in v_i is shifted to left by one position. Then, the hop count in the received trail message is increased by one, and replaces the right-most element $e_i{}^{d_v}$ in v_i. After n_i updated its trail reference, this trail message is rebroadcasted with the same sequence number and an incremented hop count. The same procedure repeats at all the other nodes in $\mathbb{N}$. Within one move of S, all nodes in the network have updated their trail references according to their hop count distances to S's trail point π_1. If a node receives a trail message with a sequence number equals to λ, but has a smaller hop count than it has already recorded, then the last hop count field in its trail reference is updated, and this trail message is rebroadcasted with the same sequence number and an incremented hop count. Trail messages that has sequence number less than λ will be discarded to eliminate flooding messages in the network. The steps described in Algorithm 6 summarizes the operations to update a trail reference. During the data gathering procedure, a node's trail reference needs to be updated every time a new trail message is received.

Algorithm 6: Trail reference update algorithm

---

while *Data gathering process is not over* **do**

  /* $------$Receive a trail message$------$ */
  ;
  if *msg.seqN* > λ **then**
    $\lambda \leftarrow$ msg.seqN ;
    Shift v_i to left by one position ;
    $e_i{}^{d_v} \leftarrow$ msg.hopC + 1 ;
    msg.hopC $\leftarrow$ msg.hopC +1 ;
    Rebroadcast message ;
  end
  else if *msg.seqN* = λ **then**
    Compare $e_i{}^{d_v}$ with (msg.hopC + 1) ;
    if $e_i{}^{d_v}$ > *(msg.hopC + 1)* **then**
      $e_i{}^{d_v} \leftarrow$ msg.hopC + 1 ;
      msg.hopC $\leftarrow$ msg.hopC +1 ;
      Rebroadcast message ;
    end
    else
      | *Discard the message* ;
    end
  end
  else if *msg.seqN* < λ **then**
    | *Discard the message* ;
  end
end
/* $------$Reset Variables$------$ */
For $j = 1, \ldots, d_v$ $e_i{}^{j} \leftarrow -1$;
$\lambda \leftarrow -1$;

After each node in the network received d_v distinct trail messages, the logical coordinate space is established. A snapshot of a part of the network $\mathbb{N}$ is shown in Fig. 4.2. Trail references, such as [3, 1, 1] or [2, 2, 2], are considered logical coordinates of the sensor nodes in a network.

4.4.1.2 Destination Identification

SinkTrail facilitates the flexible and convenient construction of a logical coordinate space. Instead of confining a mobile sink's movement, it allows a mobile sink to spontaneously stop at convenient locations according to current field situations or desired moving paths. These sojourn places of a mobile sink, named trail points in SinkTrail, are footprints left by a mobile sink, and they provide valuable information for tracing the current location of a mobile sink. Considering these footprints as virtual landmarks, hop count information reflects the moving trajectory of a mobile sink. A logical d_v-dimensional coordinate space is then established.

One advantage of SinkTrail is that the logical coordinate of a mobile sink keeps invariant at each trail point, given the continuous update of trail references. This is because the mobile sink's hop count distance to its previous $d_v - 1$ footprints are always $K(d_v - 1)$, $K(d_v - 2), \ldots, K$, and 0 to its current location. Therefore the logical coordinate $[K(d_v - 1), K(d_v - 2), \ldots, 0]$ represents the current logical location of the mobile sink. We call this coordinate "*Destination Reference*". This destination reference does not necessarily require a mobile sink to have linear moving trajectory. Although arbitrary movement of a mobile sink may deteriorate the accuracy of destination reference, it can still serve as a guideline for data reporting. Here we set $K = 1$ and $d_v = 3$ to ease our presentation. A large value of K means even less broadcast frequency. In Fig. 4.2, assume S is at the trail point 3 now, then its destination reference should be $[2, 1, 0]$. When S moves to the trail point 4, the coordinate space is updated based on trail points 2, 3, and 4, and destination reference of the mobile sink is still $[2, 1, 0]$.

4.4.1.3 Greedy Forwarding

Once a node has updated the 3 elements in its trail reference (we use $d_v = 3$ for easy understanding and clear presentation), it starts a timer that is inverse proportional to the right-most element in its trail reference. For example, node n_5's trail reference is $[6, 7, 9]$ in Fig. 4.2, then the duration of its timer is set to $T_5 = T_{init} - \mu \times 9$. Here, T_{init} and μ are predefined constants. The choice of timer function, T_{init}, and μ may vary. However, we assume the timer durations are significantly longer than the propagation time of a trail message, so that timers on all nodes are viewed as starting at the same time. The timer mechanism is mainly used to differentiate data reporting orders (another usage is discussed in SinkTrail-S protocol); so the clock on each sensor node doesn't need to be perfectly synchronized. Since the right-most element in a node's trail reference is the latest hop count information from this node to a mobile sink, the inverse proportional timers ensure that nodes faraway from S have shorter timer durations than those close to S, thus will start data reporting first. When a node's timer expires, it initiates the data reporting process.

Every sensor node in the network maintains a routing table of size $O(b)$ consisting of all neighbors' trail references. This routing table is built up by exchanging trail references with neighbors, as described in Algorithm 7; and it is updated whenever the mobile sink arrives at a new trail point. Although trail references may not be global identifiers since route selection is conducted locally, they are good enough for the SinkTrail protocol. Because each trail reference has only 3 numbers, the size of exchange message is small. When a node has received all its neighbors' trail references, it calculates their distances to the destination reference, $[2, 1, 0]$, according to *2-norm* vector calculation, then greedily chooses the node with the smallest distance as next hop to relay data. If there is a tie the next hop node can be randomly selected. The complete procedure of greedy forwarding is presented in Algorithm 7. Take the network in Fig. 4.2 as an example, when node n_8 decides to report its data, it compares n_4, n_5, and n_7's vector distance with $[2, 1, 0]$. Since n_5

Algorithm 7: Greedy data forwarding algorithm

```
/* - - - - - -Start a timer- - - - - -                                    */
```
if *All elements of the trail reference are updated* **then**
| *Start timer* $T_i = T_{init} - \mu \times e_i^{d_v}$;
| *Exchange trail references with neighbors*
end
```
/* - - - - - -When timer expires- - - - - -                               */
```
Set destination as $[(d_v - 1), \ldots, 2, 1, 0]$;
```
/* - - - - - -Probe mobile sink- - - - - -                                */
```
if *A mobile sink is within radio range* **then**
| *Send data to the mobile sink directly* ;
end
else
| *Choose the neighbor closest to destination as the next hop* ;
| *Forward all data to next hop* ;
end

---

and n_7's distance to $[2, 1, 0]$ is $\sqrt{133}$ and $\sqrt{249}$ respectively, and n_4's distance is $\sqrt{90}$, n_4 is chosen as the next hop of n_8.

4.4.2 SinkTrail Protocol with Multiple Mobile Sinks

The proposed SinkTrail protocol can be readily extended to multi-sink scenario with small modifications. When there is more than one sink in a network, each mobile sink broadcasts trail messages following Algorithm 5. Different from one sink scenario, a sender ID field, *msg.sID*, is added to each trail message to distinguish them from different senders.

Algorithms executed on the sensor node side should be modified to accommodate multi-sink scenario as well. Instead of using only one trail reference, a sensor node maintains multiple trail references that each corresponds to a different mobile sink at the same time. Figure 4.3 shows an example of two mobile sinks. Two trail references, colored in black and red, coexist in the same sensor node. In this way, multiple logical coordinate spaces are constructed concurrently, one for each mobile sink. When a trail message arrives, a sensor node checks the mobile sink's ID in the message to determine if it is necessary to create a new trail reference. The procedure is summarized in Algorithm 8. In SinkTrail trail references of each node represent node locations in different logical coordinate spaces, when it comes to data forwarding, because reporting to any mobile sink is valid, the node can choose the neighbor closest to a mobile sink in any coordinate space. Sink location in each logical coordinate space is still $[2, 1, 0]$, as we use $K = 1$, $d_v = 3$. If each mobile sink has a different K value, sensor nodes will calculate neighbors' distances to multiple destination references and select route accordingly. Detailed description is in Algorithm 9. It is well-known that geographic routing and logical coordinate based routing ensure loop-free routes [8, 29], so does SinkTrail.

Figure 4.3 gives us an example of data gathering in multiple coordinate spaces. For node n_5, its neighbor node n_2's vector distance to [2, 1, 0] with regard to the mobile sink on the left is 2, and $\sqrt{43}$ to the right mobile sink. And all other neighbors of n_5 has larger vector distance to the two sinks. So n_2 is used as the next hop to route to the red mobile sink.

4.4.3 SinkTrail-S Protocol

In SinkTrail, flooding trail messages to the whole network can be nontrivial in terms of energy consumption. To further optimize the energy usage and eliminate unnecessary control messages in the network, we propose SinkTrail-S algorithm as an improvement to the original SinkTrail. SinkTrail-S algorithm is mainly based on the following two observations. First, in a large-scale sensor network, the sensor nodes that are far away from a mobile sink may not be significantly affected by a single movement of the mobile sink. Take the sensor network shown in Fig. 4.4 as an example, when the mobile sink moves from trail point A to trail point B, the yellow sensor node at the left bottom corner may still have the same hop count distance to the mobile sink, and the routing path chosen from last *"move"* of the mobile sink may still be valid. In this case, the trail messages can be suppressed with high probability. Second, when a node has finished data reporting and forwarding, trail reference updating becomes meaningless and results in huge waste of energy, especially for peripheral sensor nodes. To properly handle these two situations, we propose a

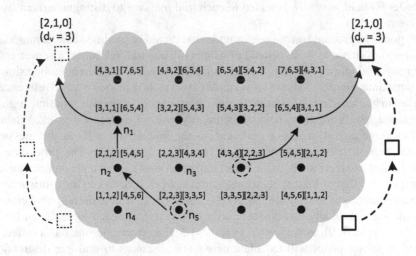

Fig. 4.3 Example execution snapshot of SinkTrail of multiple mobile sinks scenario

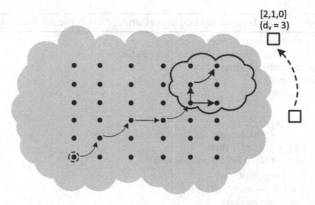

Fig. 4.4 An illustrative example to show that mobile sink's movement has less impact on remote sensor nodes than immediate ones

message suppression policy at a small cost of extra state storage at each sensor node. Each sensor node will compare the current hop count distance to a mobile sink with the most recently received one. If these two are same, it indicates the path length through the node to the mobile sink is still same, making it unnecessary to rebroadcast this trail message. In case of the second situation, each node maintains a state variable in its memory. When a node finishes data reporting, it marks itself as *"finished"*, and informs all its neighbor nodes. A node stops trail reference updating and trail message rebroadcasting whenever itself and all its neighbors are *"finished"*. Again, this method is guaranteed by the timer mechanism that ensures sequential data packets reporting order from network peripheral to a mobile sink's current location. For accidental situations due to timer failure, a new data packet may arrive at a node that has already stopped trail reference updating. In that case old trail references are used. This may cause a longer routing path but the result is still acceptable for data reporting. Algorithm 10 presents a detail description of SinkTrail-S protocol.

4.5 Performance Evaluation

Before we proceed the following variables are defined for clear presentation and fair comparison. We consider a network $\mathbb{N}$ that consists of N sensor nodes and M mobile sinks. All the sensor nodes are data sources. We assume sensor nodes are deployed in a grid topology for ease of understanding. However, our analysis can be extended to other uniformly distributed topology. Therefore, the edge of the grid is roughly $\sqrt{N}$. Denote the energy cost for transmitting or receiving a control message be α,[1] and the cost for a data packet be β. We have $\beta >> \alpha$ since compared to trail messages, data packets are usually larger in terms of data size, which is proportional to the energy cost for radio transmission.

---

[1] In practice, energy cost for transmitting and receiving might be slightly different

---

Algorithm 8: Trail reference update algorithm for multiple mobile sinks

---

while *Data gathering process is not over* **do**

    /* - - - - - -Receive a trail message- - - - - -   */

    if *New mobile sink ID* **then**

        | Create v_i_mID ;

        | Create λ_mID ;

    end

    else

        /* - - - - - -Message from a known sink- - - - - -   */

        if *msg.seqN* $> \lambda$_*mID* **then**

            | *Shift v_i_mID to left by one position* ;

            | $e_i{}^{d_v} \leftarrow$ msg.hopC $+ 1$;

            | msg.hopC $\leftarrow$ msg.hopC $+1$;

            | *Rebroadcast message* ;

        end

        else if *msg.seqN* $= \lambda$_*mID* **then**

            | *Compare $e_i{}^{d_v}$ with* (msg.hopC $+ 1$) ;

            if $e_i{}^{d_v} >$ *msg.hopC* $+ 1$ **then**

                | $e_i{}^{d_v} \leftarrow$ msg.hopC $+ 1$;

                | msg.hopC $\leftarrow$ msg.hopC $+1$;

                | *Rebroadcast message* ;

            end

            else

                | *Discard the message* ;

            end

        end

        else if *msg.seqN* $< \lambda$_*mID* **then**

            | *Discard the message* ;

        end

    end

end

/* - - - - - -Reset variables- - - - - -   */

v_i_prototype $\leftarrow [-1, -1, \ldots, -1]$ of size d_v ;

λ_prototype $\leftarrow -1$; ;

4.5.1 Privacy Protection

The privacy protection for both data source and mobile sinks are guaranteed by using logical coordinates instead of real geographic locations. As mobile sinks are moving around in the deployed area, all sensor nodes need to know is which one of its neighboring nodes is closer to the mobile sink. No real geographic location is required and no information beyond two hops is required. For an adversary with only location view of the network, the next hops of snooped packets are frequently changed. Therefore,

---

Algorithm 9: Greedy data forwarding algorithm for multiple mobile sinks

---

```
/* ------Start a timer------                                              */
```
if *All elements of the trail reference are updated* **then**
  | Start timer $T_i = T_{init} - \mu \times e_i{}^{d_v}$;
  | *Exchange trail references with neighbors* ;
end
```
/* ------When timer expires------                                         */
```
Set destination as $[(d_v - 1), \ldots, 2, 1, 0]$;
```
/* ------Probe mobile sink------                                          */
```
if *A mobile sink is within radio range* **then**
  | *Send data to the mobile sink directly* ;
end
else
  | *Compare neighbors' trail references with destination reference in already established*
  | *logical coordinates* ;
  | *Choose the neighbor closest to any mobile sink as the next hop* ;
  | *Forward all data to next hop* ;
end

---

mobile sinks' random movements inside an area provide sufficient obfuscation of their location. With the logical coordinate system, further protection is added.

4.5.2 Communication Cost Analysis

There are many mobile sink oriented approaches for data collection in sensor networks, e.g., Directed Diffusion [11], TTDD [36], and GRAB [37]. These protocols, as in SinkTrail, do not pose any constraint on a mobile sink's movement, nor do they require any special setup phase, generaly referred to as Sink Oriented Data Dissemination approaches (SODD). Although SODD approaches may apply different aggregation functions for better performance, similar strategies can be applied to SinkTrail as well.

In order to gain more insights on the energy efficiency of SinkTrail, and to demonstrate the advantage of incorporating sink location tracking, we compare the overall energy consumption of SinkTrail with these protocols. Simulation results for SinkTrail-S are also presented to show further improved performance.

In SinkTrail, energy consumption mainly includes data packet forwarding cost, E_{data}, routing table maintenance cost, $E_{routing}$, and trail message transmission cost, E_{trail}.

Two factors affect the energy cost of data forwarding: number of data packets and the average route length. The number of data packets is determined by the number of data sources in a network, in this case, N. The average route length, on the other hand, may vary depending on the locations a mobile sink has traveled. We estimate an upper bound of the average route length by considering the situation that a mobile sink appears randomly at a location inside the deployed field. In this case, we can

---

Algorithm 10: Trail reference update with message suppression

---

while *Data gathering process is not over* **do**

    `/* ------Receive a trail message------                              */`

    if *msg.seqN* $> \lambda$ **then**

        $\lambda \leftarrow$ msg.seqN ;

        if $e_i{}^{d_v} = msg.hopC + 1$ **then**

            | *Discard the message* ;

        end

        else

            Shift v_i to left by one position ;

            $e_i{}^{d_v} \leftarrow$ msg.hopC + 1 ;

            msg.hopC $\leftarrow$ msg.hopC +1 ;

            Rebroadcast message ;

        end

    end

    else if *msg.seqN* $= \lambda$ **then**

        Compare $e_i{}^{d_v}$ with (msg.hopC + 1) ;

        if $e_i{}^{d_v} > (msg.hopC + 1)$ **then**

            $e_i{}^{d_v} \leftarrow$ msg.hopC + 1 ;

            msg.hopC $\leftarrow$ msg.hopC +1 ;

            Rebroadcast message ;

        end

        else

            | *Discard the message* ;

        end

    end

    else if *msg.seqN* $< \lambda$ **then**

        | *Discard the message* ;

    end

end

`/* ------Reset Variables------                                      */`

For $j = 1, \ldots, d_v\ e_i{}^j \leftarrow -1$;

$\lambda \leftarrow -1$;

---

find $\frac{N}{2}$ pairs of sensor nodes that any one pair of nodes' distances to the mobile sink added up to at most $\sqrt{2N}$. Thus, the average route length should be upper bounded by $\frac{N}{2} \cdot \sqrt{2N}/N$. We use a coefficient c, where $0 < c \leq \frac{1}{2}$, to describe the average route length. Hence, we have,

$$E_{data} = \beta \cdot c \cdot \sqrt{2N} \cdot N \tag{4.1}$$

This energy cost upper bound for data reporting won't be affected by the number of mobile sinks, since every data reporting message will travel through the shortest possible path. Increased number of mobile sink will only decrease the total energy cost for data reporting.

According to SinkTrail protocol, the total number of trail messages depends on the network size, N, the number of trail points visited by each mobile sink, D_π, and the number of mobile sinks, M. The energy consumption for trail message transmission is given by:

$$E_{trail} = \alpha \cdot M \cdot N \cdot D_\pi \tag{4.2}$$

In SinkTrail, the energy consumption for each node to maintain local routing information is linearly proportional to the number of its neighbors, denoted by b. If there are multiple mobile sinks, the energy consumption increases as each node keeps a different trail reference for each mobile sink. Because of the broadcast nature of wireless media, this type of control message only needs to be transmitted once by each sensor node. Therefore, the energy cost for routing information maintenance is summarized by:

$$E_{routing} = \alpha \cdot N \cdot M \tag{4.3}$$

The overall energy consumption of SinkTrail protocol is:

$$E_{ST} = \beta \cdot c \cdot \sqrt{2N} \cdot N + \alpha \cdot M \cdot N \cdot D_\pi + \alpha \cdot N \cdot M \tag{4.4}$$

TTDD [36] is a well-known data dissemination protocol that uses mobile sinks. Since in TTDD, data collection requests are initiated by mobile sinks, it in fact belongs to the category of SODD approach. The basic data gathering procedure of TTDD is preceded by a data grid setup phase, where each data source propagates some descriptive information about its data to the whole network in order to construct a grid structure for guiding the forward direction of query messages. When a mobile sink queries for certain data, this query message is only flooded within a grid cell, then the pre-constructed data grid structure will help the query find the corresponding data source. According to this description, the total energy consumption of TTDD includes the following three components: energy used for grid constructing, E_{grid}; query flooding, E_{query}; and data reporting, $E_{data'}$.

The energy cost for data reporting in TTDD is determined by amount of data packets and length of routing paths. Since in TTDD, data packets are routed towards a mobile sink that appears randomly in the deployed field, the average route length is similar to SinkTrail. Therefore, we have,

$$E_{data'} = \beta \cdot c \cdot \sqrt{2N} \cdot N \tag{4.5}$$

According to TTDD protocol, the whole deployed area is divided into small cells. A query for data is only flooded inside one cell. However, as we are considering a data collection process that aims at getting all sensed data in the network, it is reasonable to argue that this single query will affect each of the data sources, thus will be propagated by all sensor nodes in the network. Therefore, we have,

$$E_{query} = \alpha \cdot M \cdot N \cdot b_{cast} \tag{4.6}$$

where b_{cast} is the number of such query broadcasts.

In the grid construction phase every data source in the network propagates a descriptive message about its data to the whole network, so that certain nodes will become anchors for a particular data source. Since every node is a potential data sources, the energy cost for this procedure is

$$E_{grid} = \alpha \cdot N \cdot N \qquad (4.7)$$

Adding the energy consumption of different tasks together, we have

$$E_{TTDD} = \beta \cdot c\sqrt{2N} \cdot N + \alpha \cdot M \cdot N \cdot b_{cast} + \alpha \cdot N \cdot N \qquad (4.8)$$

Comparing (4.4) and (4.8), we can see that one of the differences lies in the second term. Since D_π represents the number of broadcasts a mobile sink makes during the data gathering procedure in SinkTrail, and b_{cast} indicates the number of times a sink initiates a query in TTDD, these two variables can be set as equal. Thereafter, the difference can be ignored here. Another difference is between the routing information exchange cost and grid construction cost. Typically, the number of mobile sinks should be significantly less than total number of sensor nodes, i.e., $M \ll N$, and $\alpha \cdot N \cdot M \ll \alpha \cdot N \cdot N$. Hence, we have $E_{ST} < E_{TTDD}$, meaning that the total energy consumption of SinkTrail protocol is less than the energy cost by TTDD under the same condition.

According to the theoretical analysis, we can see that TTDD's special grid setup phase facilitates a mobile sink to quickly collect a small amount of data in the network, unsuitable for collecting data from all sensors. The energy consumption of TTDD to collect all data includes energy consumption of the generalized SODD method plus a constant grid setup cost.

4.5.3 Simulation Results

We conducted extensive simulations for different scenarios using TOSSIM [17] to compare the performance of Sinktrail and SODD. SODD is implemented following the abstraction described in [36]. An extension to TOSSIM is implemented to allow us simulate the sink mobility following predefined moving patterns. Simulation of TTDD is omitted for the following reasons. First, TTDD falls into the category of general SODD approach, hence, generalized SODD approach can capture the main features of TTDD. Moreover, after the theoretical analysis, we can see that the grid setup phase in TTDD is targeted at querying specific data, and not very suitable to our data collection scenario.

Note that SinkTrail-S is not included in this set of comparison. Another set of simulation results will be presented later to investigate the effectiveness of message suppression.

In the SODD approach, whenever a mobile sink moves to a different location, it broadcasts its current position to the whole network. As the message propagates a

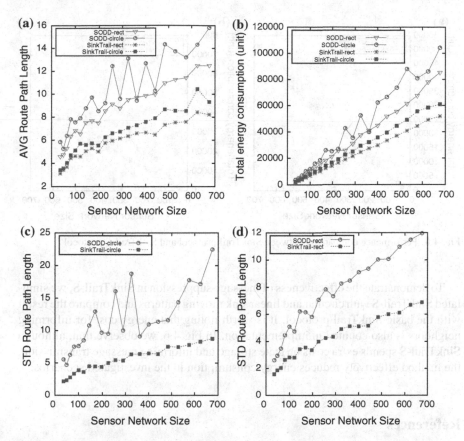

Fig. 4.5 Performance comparison between SODD and SinkTrail: **a** Average routing path length; **b** Total energy consumption; **c** Route length variances: circular sink movement; **d** Route length variances: rectangular sink movement

routing tree is established. Each node reports back its sensed data to parent node and finally, all data are merged at the root. This SODD approach suffers from losing track of the sink when location update is infrequent. To ensure fair comparison, a broadcast frequency higher than typically required by SinkTrail is used to ensure proper termination of SODD. We use one mobile sink in this set of simulations. The mobile sink moves in a rectangular or circular fashion in both algorithms. We set the data gathering threshold to 98 %. From Fig. 4.5a and b we observe that SinkTrail protocol outperforms SODD for every experimental network size. The energy consumption saving is on average 35.06 % with a route length deduction of 33.80 % for one sink SinkTrail. Figure 4.5c and d show the standard deviation of the route lengths generated by SODD and SinkTrail protocols. As you can see, SinkTrail protocol results in a more even routing path length distribution throughout the network. All these results validate the conclusion that SinkTrail helps a mobile sink to achieve energy efficient data gathering in wireless sensor networks.

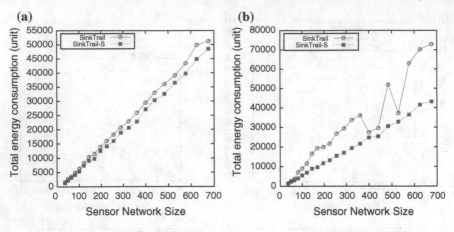

Fig. 4.6 Performance comparison between SinkTrail protocol and SinkTrail-S protocol

To demonstrate the effectiveness of message suppression in SinkTrail-S, we simulated SinkTrail-S with circular and linear sink moving patterns and compare the result with the basic SinkTrail protocol. It is worth noting that energy cost for informing neighbors is also counted in implementation. In Fig. 4.6, we observe that, although SinkTrail-S spends extra costs on state storage and informing message transmission, the method effectively reduces energy consumption in the investigated scenarios.

References

1. Bamba, B., Liu, L., Pesti, P., Wang, T.: Supporting anonymous location queries in mobile environments with privacygrid. In: Proceedings of the 17th international conference on World Wide Web (WWW), pp. 237–246. ACM (2008)
2. Chou, C.H., Ssu, K.F., Jiau, H.C., Wang, W.T., Wang, C.: A dead-end free topology maintenance protocol for geographic forwarding in wireless sensor networks. IEEE Trans. Comput. **60**, 1610–1621 (2010)
3. Coffin, D., Van Hook, D., McGarry, S., Kolek, S.: Declarative ad-hoc sensor networking. In: Proceedings of SPIE, vol. 4126, p. 109 (2000)
4. Demirbas, M., Soysal, O., Tosun, A.: Data salmon: A greedy mobile basestation protocol for efficient data collection in wireless sensor networks. Distributed Computing in Sensor Systems, pp. 267–280 (2007)
5. Deng, J., Han, R., Mishra, S.: Enhancing base station security in wireless sensor networks. Department of Computer Science, University of Colorado, Technical Report CU-CS-951-03 (2003)
6. Deng, J., Han, R., Mishra, S.: Decorrelating wireless sensor network traffic to inhibit traffic analysis attacks. Pervasive Mobile Comput. **2**(2), 159–186 (2006)
7. Fodor, K., Vidács, A.: Efficient routing to mobile sinks in wireless sensor networks. In: Proceedings of the International Conference on Wireless Internet (WICON), pp. 1–7 (2007)
8. Fonseca, R., Ratnasamy, S., Zhao, J., Ee, C.T., Culler, D., Shenker, S., Stoica, I.: Beacon vector routing: Scalable point-to-point routing in wireless sensornets. In: Proceedings of NSDI, pp. 329–342 (2005)

9. Ghinita, G., Kalnis, P., Khoshgozaran, A., Shahabi, C., Tan, K.L.: Private queries in location based services: anonymizers are not necessary. In: Proceedings of the 2008 ACM SIGMOD international conference on Management of data, pp. 121–132. ACM (2008)
10. Huang, Q., Lu, C., Roman, G.: Spatiotemporal multicast in sensor networks. In: Proceedings of the ACM International Conference on Embedded Networked Sensor Systems (SenSys), p. 217. ACM (2003)
11. Intanagonwiwat, C., Govindan, R., Estrin, D.: Directed diffusion: A scalable and robust communication paradigm for sensor networks. In: Proceedings of the International Conference on Mobile Computing and Networking (MobiCom), pp. 56–67. ACM (2000)
12. Jian, Y., Chen, S., Zhang, Z., Zhang, L.: Protecting receiver-location privacy in wireless sensor networks. In: Proceedings of the IEEE International Conference on Computer Communications (INFOCOM), pp. 1955–1963. IEEE (2007)
13. Kamat, P., Zhang, Y., Trappe, W., Ozturk, C.: Enhancing source-location privacy in sensor network routing. In: Proceedings of 25th IEEE International Conference on Distributed Computing Systems (ICDCS), pp. 599–608. IEEE (2005)
14. Karlof, C., Sastry, N., Wagner, D.: Tinysec: a link layer security architecture for wireless sensor networks. In: Proceedings of the 2nd international conference on Embedded networked sensor systems, pp. 162–175. ACM (2004)
15. Karlof, C., Wagner, D.: Secure routing in wireless sensor networks: attacks and countermeasures. Ad hoc Netw. 1(2), 293–315 (2003)
16. Keally, M., Zhou, G., Xing, G.: Sidewinder: A predictive data forwarding protocol for mobile wireless sensor networks. In: Proceedings of the Annual IEEE Communications Society Conference on Sensor, Mesh and Ad Hoc Communications and Networks (SECON), pp. 1–9 (2009). doi: 10.1109/SAHCN.2009.5168972
17. Levis, P., Lee, N., Welsh, M., Culler, D.: Tossim: accurate and scalable simulation of entire tinyos applications. In: Proceedings of the 1th ACM Conference on Embedded Networked Sensor Systems (SenSys), pp. 126–137 (2003)
18. Li, Z., Wang, N., Franzen, A., Taher, P., Godsey, C., Zhang, H., Li, X.: Practical deployment of an in-field soil property wireless sensor network. Comput. Stand. Interfaces (2011)
19. Liu, B., Ke, W., Tsai, C., Tsai, M.: Constructing a message-pruning tree with minimum cost for tracking moving objects in wireless sensor networks is np-complete and an enhanced data aggregation structure. IEEE Trans. Comput. 849–863 (2008)
20. Luo, J., Hubaux, J.P.: Joint mobility and routing for lifetime elongation in wireless sensor networks. In: Proceedings of the IEEE International Conference on Computer Communications (INFOCOM), vol. 3 (2005)
21. Ma, M., Yang, Y.: Data gathering in wireless sensor networks with mobile collectors. In: Proceedings of the IEEE International Symposium on Parallel and Distributed Processing (IPDPS), pp. 1–9 (2008). doi: 10.1109/IPDPS.2008.4536269
22. Mainwaring, A., Culler, D., Polastre, J., Szewczyk, R., Anderson, J.: Wireless sensor networks for habitat monitoring. In: Proceedings of the ACM International Workshop on Wireless Sensor Networks and Applications (WSNA), pp. 88–97. ACM, New York, NY, USA (2002)
23. Mehta, K., Liu, D., Wright, M.: Location privacy in sensor networks against a global eavesdropper. In: IEEE International Conference on Network Protocols, 2007. ICNP 2007, pp. 314–323. IEEE (2007)
24. Mehta, K., Liu, D., Wright, M.: Protecting location privacy in sensor networks against a global eavesdropper. IEEE Trans. Mobile Comput. 11(2), 320–336 (2012)
25. Moscibroda, T., O'Dell, R., Wattenhofer, M., Wattenhofer, R.: Virtual coordinates for ad hoc and sensor networks. In: Proceedings of the Joint Workshop on Foundations of Mobile Computing (DIALM-POMC), pp. 8–16. ACM, New York, NY, USA (2004)
26. Ouyang, Y., Le, Z., Chen, G., Ford, J., Makedon, F.: Entrapping adversaries for source protection in sensor networks. In: Proceedings of the 2006 International Symposium on on World of Wireless, Mobile and Multimedia Networks, pp. 23–34. IEEE Computer Society (2006)
27. Ozturk, C., Zhang, Y., Trappe, W.: Source-location privacy in energy-constrained sensor network routing. In: Workshop on Security of ad hoc and Sensor Networks: Proceedings of the 2nd ACM workshop on Security of ad hoc and sensor, networks, 25, pp. 88–93 (2004)

28. Park, T., Kim, D., Jang, S., Yoo, S., Lee, Y.: Energy efficient and seamless data collection with mobile sinks in massive sensor networks. In: Proceedings of the IEEE International Symposium on Parallel and Distributed Processing (IPDPS), pp. 1–8 (2009)

29. Rao, A., Ratnasamy, S., Papadimitriou, C., Shenker, S., Stoica, I.: Geographic routing without location information. In: Proceedings of the Annual International Conference on Mobile Computing and Networking (MobiCom), pp. 96–108. ACM New York, NY, USA (2003)

30. Saponas, T.S., Lester, J., Hartung, C., Agarwal, S., Kohno, T., et al.: Devices that tell on you: privacy trends in consumer ubiquitous computing. In: Usenix. Security vol. 3, p. 3 (2007)

31. Shah, D., Shakkottai, S.: Oblivious routing with mobile fusion centers over a sensor network. In: Proceedings of the IEEE International Conference on Computer Communications (INFOCOM), pp. 1541–1549 (2007)

32. Somasundara, A.A., Ramamoorthy, A., Srivastava, M.B.: Mobile element scheduling for efficient data collection in wireless sensor networks with dynamic deadlines. In: Proceedings of the 25th IEEE International Real-Time Systems Symposium (RTSS), pp. 296–305. IEEE Computer Society, Washington, DC, USA (2004)

33. Soysal, O., Demirbas, M.: Data Spider: A Resilient Mobile Basestation Protocol for Efficient Data Collection in Wireless Sensor Networks. In: Proceedings of the International Conference on Distributed Computing in Sensor Systems (DCOSS). Santa Barbara, California, USA (2010)

34. Srinivasan, V., Stankovic, J., Whitehouse, K.: Protecting your daily in-home activity information from a wireless snooping attack. In: Proceedings of the 10th international conference on Ubiquitous computing, pp. 202–211. ACM (2008)

35. Yang, Y., Shao, M., Zhu, S., Urgaonkar, B., Cao, G.: Towards event source unobservability with minimum network traffic in sensor networks. In: Proceedings of the ACM Conference on Wireless Network Security (WiSec). Citeseer (2008)

36. Ye, F., Luo, H., Cheng, J., Lu, S., Zhang, L.: A two-tier data dissemination model for large-scale wireless sensor networks. In: Proceedings of the Annual International Conference on Mobile Computing and Networking (MobiCom), pp. 148–159. ACM, New York, NY, USA (2002)

37. Ye, F., Zhong, G., Lu, S., Zhang, L.: Gradient broadcast: a robust data delivery protocol for large scale sensor networks. Wireless Netw. **11**(3), 285–298 (2005)

38. Yu, L., Wang, N., Meng, X.: Real-time forest fire detection with wireless sensor networks. In: Proceedings of the International Conference on Wireless Communications, Networking and Mobile Computing (IWCMC), vol. 2, pp. 1214–1217. IEEE (2005)

39. Zhao, M., Ma, M., Yang, Y.: Efficient data gathering with mobile collectors and space-division multiple access technique in wireless sensor networks. IEEE Trans. Comput. (2010)

40. Zhao, M., Ma, M., Yang, Y.: Mobile data gathering with space-division multiple access in wireless sensor networks. In: Proceedings of the IEEE International Conference on Computer Communications (INFOCOM), pp. 1283–1291 (2008)

41. Zhao, M., Yang, Y.: Bounded relay hop mobile data gathering in wireless sensor networks. IEEE Trans. Comput. (2010)

Chapter 5
Conclusion and Future Directions

5.1 Summary

With the increased popularity of mobile devices, more and more applications on the mobile platform rely on the contextual information of a user to provide high quality and customized services. In this book, we have presented several privacy preserving techniques that are specifically designed for such applications in mobile network environment.

In order to achieve privacy protection while incurring minimal cost to users, we first assumed that a trusted central authority is available. Based on this assumption, we leveraged the concept of mix zone, and studied the problem of optimally deploying multiple mix zones to achieve better privacy protection for mobile users. To formally investigate this problem, we modeled the area covered by location-based services as a graph, where all vertices (POIs) are considered as candidates for mix zone deployment. We defined a new privacy metric that quantifies the degree of privacy by measuring the number of pairwise location associations in an area. To achieve maximum privacy preservation, we formulated the optimization problem with the objective of maximizing the overall discontinuity of all possible trajectories on the road network and subject to deployment cost, traffic density, and differentiated privacy priority constraints. For each road segment and intersection, the traffic density effect in terms of entropy is also taken into account. We designed three heuristic algorithms corresponding to different traffic scenarios and privacy preservation levels as practical and efficient solutions to the NP-hard optimization problem. Simulation results based on real-world mobility trace demonstrated that our approach effectively reduced the privacy risks of the mobile users.

Since a trusted central authority might not be always available, we pursued a distributed approach. Instead of relying on a central authority to help mobile users change their pseudonyms, we let the users generate dummy users according to their own privacy requirements. These generated dummy users can help achieve location anonymization and obfuscation. Since dummy user generation is costly to a resource-constraint mobile device, we formulated a game-theoretic model to study the behaviors of mobile users. We further identified the equilibrium states of our model. Our novel game-theoretical model is suitable for user-centric fine-grained privacy preserving approach in mobile networks. These results are expected to serve as guidelines for designing incentive mechanisms and strategy optimization algorithms for mobile users.

Finally, we investigated the problem of privacy protection during information retrieval from deployed wireless sensor networks. As for a large-scale sensor network, the problem of privacy protection includes the protection of the correlation between data and the corresponding geographic region, as well as the protection of the location privacy of mobile users that act as data sinks to the sensor network. We proposed an energy efficient, mobile sink based data collection protocol that utilized the logical coordinates of both sink and sensor nodes, instead of real location information. Our proposed protocol possesses both the desirable features of geographic greedy routing, and location protection. Simulation results showed that our protocol has satisfactory performances.

5.2 Future Directions

With the fast advances of new technologies, sharing of personal information becomes increasingly simplified. The results of the techniques studied in this book, with both their limitations and their promises, indicate that it is possible to design effective communication protocols and algorithms that can help users protect their privacy. However, there are many research directions waiting to be explored:

- Since new applications for mobile devices are emerging at fast pace, new types of threat will appear. As a result, the types of information that is available to the adversarial application will change. Hence, it would be necessary to extend the inference attack model to incorporate spatial/temporal correlations of other types of information. In particular, analytic models that accurately captures the strategy of an adversary and their cost would be very helpful in devising new privacy protection techniques.
- We discussed how privacy threats depend on the context of mobile devices and the information shared by mobile devices. In order to take this into account in the design of privacy-preserving mechanisms, we used tools from other disciplines to better model different contexts. Our results show that such an approach can positively affect the design of privacy protocols. This is still a relatively untouched, yet burgeoning area of research that could be further explored.
- With the massive deployment of mobile sensing devices, huge amount of heterogeneous data are generated at unprecedented scale and complexity. In order to analyze, visualize, and extract intelligent information from these data, many research efforts have been focusing on utilizing the cloud as computational brain to build up intelligent and responsive mobile distributed systems. This opens up new directions for security and privacy researches. On the one hand, since cloud

services providers are not guaranteed to be trustworthy, new types of threats may emerge due to the fact that large amount of sensitive information are transferred to third parties. On the other hand, as cloud servers can be served as a computational brain to the less powerful mobile devices, many advanced encryption technologies can be employed for enhanced protection. How to effectively utilize cloud services for better performance and better protection is an area to be explored.